T0256236

Seismic Resilience Assessment of Hospital Infrastructure

Healthcare facilities or hospital systems are classified as some of the most critical infrastructure systems when responding to natural disasters. *Seismic Resilience Assessment of Hospital Infrastructure* systematically presents a suite of novel techniques developed by the authors and their team for seismic resilience assessment of hospital infrastructure, with particular emphasis on seismic tests and fragility models of hospital equipment, resilience assessment of single hospital buildings and emergency departments, and post-earthquake functionality of urban hospital infrastructures.

Features:

- Presents a state-of-the-art review on hospital resilience
- Develops seismic fragility model database for hospital equipment based on shaking table tests
- Provides a road map for effective and efficient methods necessary for assessing and improving seismic resilience of hospital systems and other critical engineering systems
- Expertly summarizes outcomes of many important research projects sponsored by various research agencies, including the National Natural Science Foundation of China

Taylor and Francis Series in Resilience and Sustainability in Civil, Mechanical, Aerospace and Manufacturing Engineering Systems

Series Editor Mohammad Noori
Cal Poly San Luis Obispo

Published Titles

Thermal and Structural Electronic Packaging Analysis for Space and Extreme Environments
Juan Cepeda-Rizo, Jeremiah Gayle, and Joshua Ravich

Endurance Time Excitation Functions
Intensifying Dynamic Loads for Seismic Analysis and Design
Homayoon E. Estekanchi and Hassan A. Vafai

Automation in Construction toward Resilience
Robotics, Smart Materials and Intelligent Systems
Edited By Ehsan Noroozinejad Farsangi, Mohammad Noori, Tony T.Y. Yang, Paulo B. Lourenço, Paolo Gardoni, Izuru Takewaki, Eleni Chatzi, and Shaofan Li

Data Driven Methods for Civil Structural Health Monitoring and Resilience
Latest Developments and Applications
Edited By Mohammad Noori, Carlo Rainieri, Marco Domaneschi, and Vasilis Sarhosis

Navigating the Complexity Across the Peace-Sustainability-Climate Security Nexus
Bernard Amadei

Seismic Resilience Assessment of Hospital Infrastructure
Qingxue Shang, Tao Wang, Jichao Li, and Mohammad Noori

Series Editor Bio: Prof. Mohammad Noori is a professor of mechanical engineering at California Polytechnic State University, San Luis Obispo. He received his BS (1977), his MS (1980) and his PhD (1984) from the University of Illinois, Oklahoma State University and the University of Virginia respectively; all degrees in Civil Engineering with a focus on Applied Mechanics. His research interests are in stochastic mechanics, non-linear random vibrations, earthquake engineering and structural health monitoring, AI-based techniques for damage detection, stochastic mechanics, and seismic isolation. He serves as the executive editor, associate editor, the technical editor or a member of editorial boards of 8 international journals. He has published over 250 refereed papers, has been an invited guest editor of over 20 technical books, has authored/co-authored 6 books, and has presented over 100 keynote and invited presentations. He is a Fellow of ASME, and has received the Japan Society for Promotion of Science Fellowship.

For more information about this series, please visit: https://www.routledge.com/Resilience-and-Sustainability-in-Civil-Mechanical-Aerospace-and-Manufacturing/book-series/ENG

Seismic Resilience Assessment of Hospital Infrastructure

Qingxue Shang, Tao Wang,
Jichao Li, and Mohammad Noori

CRC Press
Taylor & Francis Group
Boca Raton London New York

CRC Press is an imprint of the
Taylor & Francis Group, an **informa** business

Designed cover image: © Shutterstock

MATLAB® is a trademark of The MathWorks, Inc. and is used with permission. The MathWorks does not warrant the accuracy of the text or exercises in this book. This book's use or discussion of MATLAB® software or related products does not constitute endorsement or sponsorship by The MathWorks of a particular pedagogical approach or particular use of the MATLAB® software.

First edition published 2024
by CRC Press
2385 NW Executive Center Drive, Unit 320, Boca Raton, FL 33431

and by CRC Press
4 Park Square, Milton Park, Abingdon, Oxon, OX14 4RN

CRC Press is an imprint of Taylor & Francis Group, LLC

© 2024 Qingxue Shang, Tao Wang, Jichao Li, and Mohammad Noori

Library of Congress CataloginginPublication Data
Names: Shang, Qingxue, author. | Wang, Tao (Researcher in earthquake engineering), author. | Li, Jichao (Structural engineer), author. | Noori, Mohammad, author.
Title: Seismic resilience assessment of hospital infrastructure / Qingxue Shang, Tao Wang, Jichao Li, and Mohammad Noori.
Description: First edition. | Boca Raton : CRC Press, 2024. | Series: Resilience and sustainability | Includes bibliographical references.
Identifiers: LCCN 2023037566 (print) | LCCN 2023037567 (ebook) | ISBN 9781032600802 (hardback) | ISBN 9781032600819 (paperback) | ISBN 9781003457459 (ebook)
Subjects: LCSH: Hospital buildings—Earthquake effects | Medical instruments and apparatus—Safety measures. | Hospitals—Safety measures. | Hospitals—Risk management. | Earthquake engineering.
Classification: LCC RA969.9. S48 2024 (print) | LCC RA969.9 (ebook) | DDC 725/.510289—dc23/eng/20231204
LC record available at https://lccn.loc.gov/2023037566
LC ebook record available at https://lccn.loc.gov/2023037567

ISBN: 978-1-032-60080-2 (hbk)
ISBN: 978-1-032-60081-9 (pbk)
ISBN: 978-1-003-45745-9 (ebk)

DOI: 10.1201/9781003457459

Typeset in Times
by codeMantra

Contents

Preface

Healthcare facilities or hospital systems are classified as one of the most critical infrastructure systems when responding to disaster events. Healthcare service accessibility is significant for citizens' health in both daily life and aftermath of disasters such as earthquakes. However, many hospitals have reportedly suffered significant damage and lost their functionality during earthquakes. For instance, the 1971 San Fernando earthquake caused severe damage to four hospitals. Following the 1994 Northridge earthquake, 8 nursing homes and 20 hospitals were identified as unsafe for occupancy or entry. Similar damage to hospital systems was observed after the 1995 Kobe earthquake. Nearly 20% of hospitals in the affected area were partially or completely inoperable after the 2010 Chile earthquake. In these earthquakes, emergency rescue was strongly affected owing to the reduced functionality of the hospital infrastructure systems, which initiated a comprehensive study to evaluate and improve the seismic performance of the hospital systems.

Among many performance indices of hospital buildings, seismic resilience, which is defined as the ability of an engineering system to resist, restore, and adapt to an earthquake's impact, has caused much concern recently. Measuring hospital resilience remains a real challenge, and much research effort is needed, particularly the quantification of the functionality and recovery process. This monograph systematically presents a suite of novel techniques developed by the authors and their team for seismic resilience assessment of hospital infrastructure, with particular emphasis on seismic tests and fragility models of hospital equipment, resilience assessment of single hospital buildings and emergency departments, and post-earthquake functionality of urban hospital infrastructures.

It is expected that the methods and technologies introduced in this monograph provide a road map for effective and efficient methods that are needed for assessing and improving the seismic resilience of hospital systems and other critical engineering systems.

The contents covered in this monograph include the outcomes of many important research projects sponsored by various research agencies, including the National Natural Science Foundation of China (51378478, 51678538, 51908519, and 52208483); the National Science Foundation for Distinguished Young Scholars (52125806); the National Key Research and Development Program of China (2016YFC0701101, 2017YFC1500700, 2019YFC1509304, and 2019YFE0112700); the International Science and Technology Cooperation Program of China (2014DFA70950); the Scientific Research Fund of Institute of Engineering Mechanics, China Earthquake Administration (2016A06, 2019EEEVL0501, 2019EEEVL0505, 2019A02, 2019B02, and 2021EEEVL0212); the China Postdoctoral Science Foundation (2021M701937); the Shuimu Tsinghua Scholar Program (2021SM005); and Heilongjiang Touyan Innovation Team Program (3016).

Qingxue Shang
Tao Wang
Jichao Li
Mohammad Noori
Beijing, China
September 2022

Acknowledgments

This research was funded by the National Science Foundation for Distinguished Young Scholars (52125806), National Natural Science Foundation of China (52208483), and Heilongjiang Touyan Innovation Team Program (3016). Any opinions, findings, and conclusions or recommendations expressed in this book are those of the authors and do not necessarily reflect the views of the sponsors.

About the Authors

Dr. Qingxue Shang obtained his bachelor's degree from Southeast University, China, in 2015. He obtained his master's and PhD degrees from the Institute of Engineering Mechanics, China Earthquake Administration, in 2018 and 2021, respectively. He worked as a postdoctoral researcher at Tsinghua University and then joined China Earthquake Disaster Prevention Center in 2023. He has got the support of the Natural Science Foundation of China Youth Fund Project, National Key Research and Development Plan Project, China Postdoctoral Science Foundation, and Shuimu Tsinghua Scholar Program. His research aims to evaluate and improve the seismic resilience of our society and community. His recent works and publications include the resilience assessment of hospital systems, power systems, and water distribution networks, and some resilience enhancement techniques including resilient shear wall structures and resilient nonstructural components. He worked as an Individual Research Member of the International Association for the Seismic Performance of Non-Structural Elements (SPONSE) and an Early Career Editor of Earthquake Engineering and Resilience. He is also an active reviewer for many international journals, including *Reliability Engineering & System Safety; Earthquake Engineering and Structural Dynamics; Sustainable Cities and Society; International Journal of Electrical Power and Energy Systems; Structures; Earthquake Engineering and Engineering Vibration; and Advances in Bridge Engineering.*

Professor Tao Wang earned his bachelor's and PhD degrees from Tsinghua University, China and Kyoto University, Japan, respectively. His research commits to solve key scientific problems in reproduction of engineering failure mechanism and resilient earthquake engineering. His research findings have facilitated seismic resilience improvement of several important engineering projects. He has established extensive relations of cooperation with many experts from different countries to jointly promote seismic resilience evaluation and promotion work. He has organized an international special issue on seismic resilience and established the Huixian International Forum on Earthquake Engineering for Young Researchers, which attracted more than 600 young scholars from 30 countries to join. He has won the Second Prize of the China National Science and Technology Progress Award and many other important awards. He is a leading talent in science and technology of the China Earthquake Administration (CEA), and recently, he has won the Outstanding Youth Fund of the National Natural Science Foundation of China.

Dr. Jichao Li obtained his bachelor's degree from Tsinghua University, China, in 2012. He obtained his PhD degree from the Institute of Engineering Mechanics (IEM), China Earthquake Administration, in 2018. He is currently an associate professor at IEM. His research focuses on seismic resilience analysis method, resilience improvement technology, and intelligent vibration control theory of engineering systems. He has got the support of the Natural Science Foundation of China Youth Fund project, Heilongjiang Provincial Natural Science Foundation joint guiding project, Basic Scientific Research Business Project of the Central Public Welfare Research Institute, and National Key Research and Development Plan.

Professor Mohammad Noori is a professor of mechanical engineering at Cal Poly, San Luis Obispo, a fellow and life member of the American Society of Mechanical Engineering, and a recipient of the Japan Society for Promotion of Science Fellowship. His work in nonlinear random vibrations, seismic isolation, and application of artificial intelligence methods for structural health monitoring is widely cited. He has authored over 300 refereed papers, including over 160 journal articles and 6 scientific books, and has edited 25 technical, and special journals, volumes. He has supervised over 90 graduate students and post-doc scholars, and has presented over 100 keynote, plenary, and invited talks. He is the founding co-chief editor of an international journal and has served on the editorial board of over 15 other journals and as a member of numerous scientific and advisory boards. He has been a distinguished visiting professor at several highly ranked global universities and directed the Sensors Program at the National Science Foundation in 2014. He has been a founding director or co-founder of three industry-university research centers and held chair professorships at two major universities. He served as the dean of engineering at Cal Poly for 5 years, has also served as the Chair of the National Committee of Mechanical Engineering Department Heads, and was one of the seven co-founders of the National Institute of Aerospace, in partnership with NASA Langley Research Center.

1 Introduction and Background

1.1 RESEARCH BACKGROUND AND SIGNIFICANCE

In the 1970s, Holling (1973) defined ecological resilience as a measure of the ability of ecological systems to absorb changes of state and bounce back from external shocks. The concept was then borrowed from ecological system to disaster resilience and has been used in different engineering systems. A well-known definition of system resilience is the ability of a system to reduce the chances of a shock, to absorb such a shock if it occurs, and to recover quickly after a shock (Bruneau et al. 2003). Resilience definition and analysis have been conducted in different systems during the last two decades, and literature reviews of different systems exist, e.g., coastal bridges, energy systems, dam engineering, infrastructure systems, community resilience, and urban resilience.

Healthcare facilities or hospital systems are classified as one of the most critical infrastructure systems when responding to disaster events. Healthcare service accessibility is significant for citizens' health in both daily life and aftermath of disasters such as wildfire, hurricanes, floods, storms, and earthquakes. During the 1993 Midwest floods, local hospitals were faced with problems of lacking municipal water and power, and many clinical services were canceled or diverted to alternate facilities (Peters 1996). Callaghan et al. (2007) claimed that interruption of healthcare service posed threats to vulnerable populations, especially for pregnant women and infants during and after the 2005 Katrina hurricanes. Sometimes hospital evacuation may be necessary to ensure the safety of patients and medical staff when the hospital is unsafe or out of service (McGinty et al. 2016). After the 2012 Hurricane Sandy, McGinty et al. (2016) identified the determinants of acute care hospital evacuation and shelter in place by interviews with hospital executives and local officials. Following 2012 Hurricane Sandy, a significant increase in the hospital admission of elderly patients aged 65 and up was found, which posed a huge demand on healthcare capacity (McQuade et al. 2018). Lack of utility access such as power systems and damage to nonstructural components significantly reduced the post-wildfire functionality of healthcare facilities during the 2018 Camp Fire in Paradise, California (Schulze et al. 2020).

An earthquake with large intensities may cause buildings to collapse and result in heavy casualties and hospital admissions, which put great pressure on local healthcare facilities (Peek-Asa et al. 1998). Therefore, post-earthquake performance of healthcare facilities is quite important to help treat injured patients (Jacques et al. 2014). However, seismic damage to structural and nonstructural components (NSCs), and medical equipment in hospital buildings (Shang et al. 2019;

DOI: 10.1201/9781003457459-1

Wang et al. 2019a, b; Achour & Miyajima 2020; Wang et al. 2021), infrastructure systems (Li et al. 2018, 2020a), and lack of supplies and medical staffs (Ochi et al. 2016) are the most common causes of a decrease in healthcare functionality.

Achour et al. (2011) investigated the seismic damage data of physical components in 34 healthcare facilities, and the results indicated that the damages of structural components were different but the equipment and utility supplies' damages were similar as a result of insufficient seismic protection. Chen et al. (2017) found that after the 2015 Nepal earthquake, hospitals in high-intensity areas were seriously damaged in both structural and nonstructural components. It is noteworthy that the loss of hospital functionality can occur after earthquakes even with minimal seismic damage to the building structures, as reported by Kirsch et al. (2010). Kirsch et al. (2010) found that the loss of telecommunication service has great influence on hospital functionality during the 2010 Chilean earthquake. Ghanjal et al. (2019) found that the lack of preparedness of the government and the lack of appropriate management of hospitals were the most important weaknesses in the provision of healthcare services after the 2017 Kermanshah earthquake in Iran.

Healthcare service provided in hospitals can be compromised due to many different internal or external causes. There is no unified approach or indicator system to identify hospital weakness and evaluate hospital resilience during and after disasters up to now. This study presents a state-of-the-art review of hospital resilience. Different definitions and quantification methods of hospital resilience were summarized first, followed by comparisons of potential resilience indicators of hospitals in the literature. Different frameworks and methods for evaluating hospital resilience are presented at the single hospital level and healthcare network level. The system modeling methods of hospital systems including discrete event simulation (DES), agent-based simulation (ABS), system dynamics modeling (SDM), and fault tree method (FTM) are then compared. The major knowledge gaps on disaster resilience of hospital systems to be filled are identified, and possible future research challenges are discussed.

1.2 RESILIENCE DEFINITION AND QUANTIFICATION OF HOSPITAL SYSTEMS

According to Bruneau et al. (2003), system resilience can be defined or measured using four dimensions, four properties, and three results as listed in Table 1.1. Based on the concept of Bruneau et al. (2003), disaster resilience contains four interrelated dimensions, namely, technical, organizational, social, and economic (TOSE) dimensions, as shown in Figure 1.1. The TOSE dimensions cannot be adequately measured using one single performance measure as different dimensions are measured by different performance measures, and the performance measures for the same dimension of different engineering systems may be also different. Robustness, redundancy, resourcefulness, and rapidity (4Rs) are the four properties used to define and quantify disaster resilience (Bruneau et al. 2003). The concept of 4Rs is simple, while the quantification of these four properties is difficult owing to the fact that each one of them is an aggregative indicator and cannot be adequately measured by any single measure of

TABLE 1.1

Definition of System Resilience (Bruneau et al. 2003)

Definition	Aspects	Interpretation
Four dimensions of system resilience	Technical dimension	The ability of systems to perform to acceptable levels when subject to disasters.
	Organizational dimension	All the management activities and plans, maintenance, and response to emergencies.
	Social dimension	All the societal effects of disasters and the mitigation measures that are taken to reduce disaster-induced negative consequences.
	Economic dimension	The capacity to reduce all the direct and indirect economic losses due to the loss of functionality and the post-disaster rehabilitation.
Four properties of system resilience	Robustness	The ability of systems to withstand disasters without suffering degradation or functionality loss and provide service.
	Rapidity	The speed with which a system meets priorities and achieves goals in a timely manner in order to reduce losses and avoid future disruption.
	Redundancy	The extent to which components of the system are substitutable.
	Resourcefulness	The capacity to allocate resources rationally to cope with disasters.
Three results of system resilience	More reliable	Reduced failure probabilities.
	Fast recovery	Reduced time to recover to acceptable level.
	Low socio-economic consequences	Reduced negative consequences from disasters.

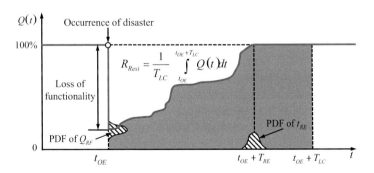

FIGURE 1.1 Four dimensions of disaster resilience of hospital systems.

performance. Performance measures of hospitals in terms of the four dimensions and four properties were also given in Bruneau et al. (2003) for illustration.

Hospital disaster resilience is illustrated in Figure 1.2, where $Q(t)$ is the hospital functionality level and is quantified using a dimensionless percentile that

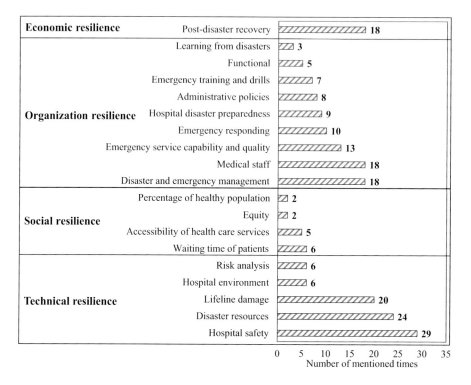

FIGURE 1.2   Schematic of hospital disaster resilience. Adapted from Shang et al. (2020a).

represents current functionality normalized by the available functionality prior to the disasters. The variation of hospital functionality $Q(t)$ over time describes the recovery process or strengthening process. When a hospital or healthcare network is affected by disasters, it will experience structural damage; nonstructural damage; and lack of power, water, and medical staffs; the hospital functionality level is compromised immediately or slowly after the disasters. During the post-disaster phases, recovery efforts of the hospital system itself will be launched, i.e., repair the disaster-damaged components, call back the off staffs, etc. In addition, the resources from outside (i.e., medical supplies, medical staffs from other healthcare facilities, support from the government, etc.) will be delivered to the damaged hospital system. Then, the functionality level of the hospital will be recovered. The recovery time T_{RE} demonstrates the recovery speed of the analyzed hospitals. The shape of the recovery curve depends on different manpower and material resources available.

Several definitions and quantification methods of resilience for healthcare facilities are presented in Table 1.2. The resilience index R_{Resi} defined in Equation (1.1) is often adopted as a resilience measure (Cimellaro et al. 2010; Jacques et al. 2014; Shang et al. 2020a, b). The 72-hour "golden window" is the best time for emergency rescue for survivors in disasters and thus is usually selected as the control time for hospital systems for emergency response. Another widely used quantification method

TABLE 1.2
Definition and Quantification of Resilience for Healthcare Facilities

Resilience Definition	Resilience Quantification	Literature Justification
Hospital resilience is defined as the ability of a hospital system to reduce the chances of a shock, to absorb such a shock if it occurs, and to recover quickly after a shock	Hospital resilience is quantified as the integral of the hospital's functionality change over time. The hospital functionality can be quantified using patients' waiting time at the individual facility level and in terms of the number of patients in queue before and after the disasters at the community level	Bruneau et al. (2003); Cimellaro et al. (2010)
Hospital resilience is the capability to absorb the impact of disasters without loss of functions; maintain its most essential functions; and recover to the pre-event state or to a new state of function	The extent of a hospital's resilience can be measured with reference to the level of hospital function, such as the number or percentage of patients assessed and treated	Zhong et al. (2013)
Hospital resilience is the ability of a hospital to manage critical event surge and continue providing healthcare service after the occurrence of an emergency, including the recovery phase	Hospital resilience is quantified as the integral of the hospital's functionality change over time. The hospital functionality is calculated as a weighted sum of the functions of different critical hospital services with different weighting factors	Jacques et al. (2014)
Hospital disaster resilience is defined as the ability of hospitals to resist, absorb, and respond to the shock of disasters while maintaining and surging essential health services, and then to recover to its original state or adapt to a new one	Hospital disaster resilience is calculated by linearly combining four factors with different weights, i.e., (1) emergency medical response capability, (2) disaster management mechanisms, (3) hospital infrastructural safety, and (4) disaster resources	Zhong et al. (2014)
Hospital Safety Index (HSI) is defined and used to evaluate the probability that a hospital will remain operational in emergencies and disasters	A total of 151 items related to structural, nonstructural, emergency and disaster management modules are evaluated and the results are linearly combined using factors with different weights	WHO and PAHO (2015)
NA	Hospital resilience is quantified by considering different phases (prevention phase, preparedness/ response phase, and recovery phase). Different indicators are used to quantify the scores of hospitals in different phases	Shirali et al. (2016)

(Continued)

TABLE 1.2 (*Continued*)
Definition and Quantification of Resilience for Healthcare Facilities

Resilience Definition	Resilience Quantification	Literature Justification
Hospital disaster resilience is defined as the ability to absorb and recover from hazardous events, containing the effects of disasters when they occur	Hospital disaster resilience is calculated by linearly combining three factors with different weights, i.e., (1) cooperation and training management, (2) resources and equipment capability, and (3) structural and organizational operating procedures	Cimellaro et al. (2018)
Hospital disaster resilience is the ability of the hospital to respond, resist, and absorb the disaster effects to provide healthcare and then return to the basic or acceptable level of service	Hospital resilience is measured based on the total number of deaths	Ramandi and Kashani (2018)
Seismic resilience of hospitals is defined as the ability to resist, restore, and adapt to an earthquake's impact	Hospital resilience is quantified as the integral of the hospital's functionality change over time. The hospital functionality is calculated as a weighted sum of 38 different components with different weighting factors	Shang et al. (2020a)

of hospital disaster resilience is based on a linear combination using weight factors (Zhong et al. 2014; WHO-PAHO 2015):

$$R_{Resi} = \frac{1}{T_{LC}} \int_{t_{OE}}^{t_{OE}+T_{LC}} Q(t)dt$$

(1.1)

where t_{OE} is the occurrence time of the disaster, and T_{LC} is the control time of the hospital.

1.3 POTENTIAL RESILIENCE INDICATORS FOR HOSPITAL INFRASTRUCTURE SYSTEMS

Both qualitative and quantitative resilience indicators can be used in hospital disaster resilience assessment. According to the Hospital Safety Index (HSI) Guide published by the World Health Organization and Pan American Health Organization (WHO-PAHO 2015), HSI is one of the best indicators of global disaster reduction. A total of 151 indicators related to structural, nonstructural, emergency, and disaster management modules are evaluated by evaluators to determine the HIS of studied hospitals. Achour et al. (2014) considered six utilities including electricity, gas supply, water supply, landline telecommunication, mobile phone, and Personal Handyphone System to evaluate the disaster resilience of healthcare facilities. Cimellaro et al. (2018) identified cooperation and training management, resources and equipment capability, and structural and organizational operating procedures as the most representative indicators of hospital disaster resilience by factor analysis. Shang et al. (2020a) divided hospital systems into seven subsystems including the structure system; the electrical system; the heating, ventilation air, and conditioning (HVAC) system; the medical system; the enclosure system; the water supply and drainage system; and the egress system. Different indicators were used to determine the resilience level of the seven subsystems. For healthcare facilities, the most recommended indicators of disaster resilience are listed in Table 1.3. Only 9.48% of the indicators are quantitative indicators, while others are qualitative indicators. And 76.78% of the indicators focus on pre-disaster evaluation and analysis.

TABLE 1.3
Disaster Resilience Indicators of Healthcare Facilities

Resilience Dimensions	Resilience Indicators	Number of Mentioned Times
Technical resilience	*Hospital safety*: Nonstructural safety, structural safety, alarm systems, security systems	29
	Disaster resources: Stuff, emergency stockpiles and supply management, materials, medication and equipment, pharmacy and medical treatment, public utility resource, number of hospital patient beds	24
	Lifeline damage: Communication system, transportation system, power system, water shortage and contamination, air condition system, medical gas system, water supply system, water treatment system, infrastructural system	20
	Hospital environment: Cleaning of hospital, environmental health	6
	Risk analysis: Monitoring, evaluation	6
Social resilience	*Waiting time of patients*: Waiting time at hospitals, travel time to hospitals	6
	Accessibility of healthcare services: Accessibility to hospitals, availability of healthcare services	5
	Equity: Healthcare equity	2
	Percentage of healthy population: Number of death and injury, percentage of healthy population	2
Organization resilience	*Disaster and emergency management*: Evacuations, organizational operating procedures, health information system, guidelines	18
	Medical staff: Human resources, personnel, number of medical staffs (doctors and nurses)	18
	Emergency service capability and quality: Hospital service quality, patient centeredness, professional competence, service attitude	13
	Emergency responding: Collaborative networks, cooperation	10

(Continued)

TABLE 1.3 (*Continued*)
Disaster Resilience Indicators of Healthcare Facilities

Resilience Dimensions	Resilience Indicators	Number of Mentioned Times
	Hospital disaster preparedness: Business continuity planning, disaster preparedness	9
	Administrative policies: Leadership and governance, strategies, public information systems	8
	Emergency training and drills: Disaster training and drills, practice and experience, training/exercise/drills related to surge capacity	7
	Functional: Functional	5
	Learning from disasters: Learning	3
Economic resilience	*Post-disaster recovery*: Recovery, evaluation, and adaptation, post-disaster recovery time, economic loss, repair cost	18

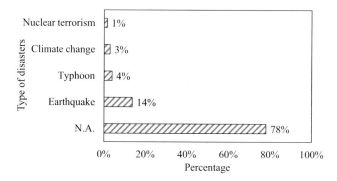

FIGURE 1.3   Indicators for different disasters.

The resilience indicators are intended to be used for typical disasters, as shown in Figure 1.3. Most of the indicators can be used for different disasters, e.g., earthquakes, floods, and typhoons (78%), while 14% focus on earthquakes, 4% focus on typhoons, 3% focus on climate changes, and 1% focus on nuclear terrorism. Quantification of hospital functionality and seismic resilience still needs to develop unified functionality indicators and resilience indicators.

1.4   RESILIENCE ASSESSMENT FRAMEWORKS OF HOSPITAL SYSTEMS

1.4.1   SINGLE HOSPITAL LEVEL

Jacques et al. (2014) assessed the functionality of critical hospital services using a fault tree method, but the fault trees were considered to be deterministic because the state of each component was based on field data. Recently, Hassan and Mahmoud

(2019) extended the fault tree developed by Jacques et al. (2014) to estimate earthquake-induced functionality reduction of a hospital building. The functionality is estimated based on hospital losses that result from its sustained damage and damage to other lifelines. Recovery curves of the investigated hospital are obtained using the continuous Markov chain process. Shang et al. (2020a) proposed a framework to quantitatively evaluate the seismic resilience of hospital systems. A typical single hospital building is categorized into different functional units, subsystems, and components. Resilience demand is expressed as the desirable recovery time of the hospital system after earthquakes. Seismic resilience is quantified based on probabilistic seismic fragility analysis. Recovery time is calculated considering an idealized repair path. Loss of functionality of the hospital is evaluated as the sum of weighted economic losses of all components.

1.4.2 Healthcare Network Level

Although the assessment of a single hospital in a community can be insightful, the overall performance of a healthcare network cannot be comprehensively evaluated unless the performance and interactions of all elements of the network are considered (Khanmohammadi et al. 2018). Dong and Frangopol (2016) developed an integrated framework for the healthcare-bridge network system performance analysis. The travel time and waiting time were used to evaluate system-level performance. The correlations among structural damages and the effect of bridge retrofit actions were considered. Fawcett and Oliveira (2000) developed a regional simulation model to analyze how a regional healthcare system responds to an earthquake event and focused on the casualty treatment problem. The numbers and locations of casualties rescued alive, the post-earthquake hospital capacity, and the damage to the transport system are selected as inputs to the model. The effects of alternative disaster response strategies on casualty treatment capacity were analyzed. Cimellaro et al. (2019) proposed a first-order method for evaluating the post-earthquake capacity of a healthcare network. The waiting time for injuries was used as the main response parameter to assess the hospital network performance. Hassan and Mahmoud (2020) integrated the quantity and quality of healthcare service, the patient demand, and the interaction with community infrastructure systems for estimating functionality and recovery of healthcare systems after earthquakes. The travel time between each node is obtained by multiplying the length of each link by the average driving speed on this link. Ceferino et al. (2020) developed a method for emergency response in hospital systems, which evaluated the loss of hospital functions and quantified the number of injuries under earthquakes. This method was then adopted to analyze the relative capacity and demand of ICUs following a severe earthquake during the COVID-19 pandemic. Shang et al. (2020c) developed a benchmark model of a medium-sized city located in the southeastern coastal region of China. They analyzed the variations of demand and capacity of the medical emergency system of the benchmark city after earthquakes. The accessibility to hospitals and the number of available beds and medical staff in hospitals were employed to quantify the functionality of the medical emergency system.

1.5 MODELING METHOD OF HOSPITAL SYSTEMS

System modeling and simulation methods have been widely used in disaster resilience analysis and evaluation of healthcare facilities (EDs, other hospital units, and networks) (Swisher et al. 2001). The widely used system modeling methods of hospital systems include discrete event simulation (DES), agent-based simulation (ABS), system dynamics modeling (SDM), and fault tree method (FTM).

1.5.1 Discrete Event Simulation (DES)

DES is the father of all simulation methods and has a long history. DES is used to model systems that change states dynamically, stochastically, in discrete intervals. DES is particularly a powerful method in systems that have a strong queueing structure (e.g., hospital EDs) since it is based on tracking entities that change state in a system. Queues are formed naturally by entities that compete for resources (Gunal 2012). A DES model includes the following: entities and attributes, resources, a network of processes, and variables (inputs and outputs). Entities in a hospital model are generally patients. Doctors, nurses, beds, and equipment can be thought as resources that cause entities to change state (Gunal 2012). DES is often used to predict the performance of hospitals under normal conditions (Duguay and Chetouane 2007; Gunal 2012; Best et al. 2014; Gul and Guneri 2015), and this method has been typically used to calibrate meta-models that simplify the system significantly; however, earthquake influences have not been considered.

Komashie and Mousavi (2005) evaluated the effects of ED resources including the number of nurses, doctors, and patient beds and patient waiting times using DES modeling. Duguay and Chetouane (2007) modeled an ED and found that the waiting times of patients did not meet the specification requirements (especially the waiting duration from registration to available examination room). Then, what-if analysis was considered to improve the performance of this ED. The results indicated that both adding examination rooms and matching increase in the medical staffs are necessary. Hoot et al. (2008) developed a DES model of an ED to evaluate the patient flow and ED crowding. Paul and Lin (2012) used the DES model to understand the roles played by different types of resources in the ED and identified that process improvement can help reduce waiting time. Best et al. (2014) compared the effects of operational interventions on patient flow using the DES model of a Ghanaian acute care hospital and identified that interventions designed to maximize the efficiency of existing resources to meet patient demand could significantly decrease patient staying time in the hospital. Kadri et al. (2014) developed a simulation-based decision support system to prevent and predict strain situations in EDs based on the DES model. AminShokravi and Heravi (2020) developed a DES model to model the operations taken in EDs and to obtain the expected time to get clinical care in EDs under normal conditions.

Cimellaro and Piqué (2014) presented a DES model of ED, and waiting time was adopted as the main parameter to evaluate the ED response and resilience after earthquakes. The ED response with and without the emergency plan was compared (Cimellaro et al. 2016). An ED network consisting of five public hospital EDs located in Istanbul was modeled using DES by Gul and Guneri (2015) to generate post-earthquake response and develop a healthcare decision- and policy-making tool for stakeholders. It is indicated that medical staffs sharing among other EDs temporarily or hiring doctors and nurses outside the hospital for a period of post-earthquake time can be useful in reducing the total patient length of stay and lengthy queues in EDs. TariVerdi et al. (2018) investigated hospital functionality under mass casualty incidents using DES. The impacts of hospital capacity-expansion strategies (modified triage tactics, early-discharge decisions, speed-up in patient care procedures, and omission of some patient care services) were analyzed. A DES model was used by Favier et al. (2018) to model the flow of patients within the different units of the ED after earthquakes. Monte Carlo simulation was integrated to analyze the ED response under different earthquakes. Results show that an event with 140 thousand hours of accumulated additional waiting time has approximately 2% probability of exceedance in 50 years. Sadegh and Hamed (2017) adopted the waiting time of patients after an earthquake to characterize the serviccability, and DES was utilized to evaluate the seismic resilience of a hospital. Shahverdi et al. (2019) developed a DES-based framework to evaluate disaster resilience of a regional healthcare network composed of five hospitals.

1.5.2 AGENT-BASED SIMULATION (ABS)

ABS is a simulation method for modeling dynamic, adaptive, and autonomous systems. It is employed to discover systems by using 'deductive' and 'inductive' reasoning. At the core of an ABS model, there are 'autonomous' and 'interacting' objects called agents. Agents are like entities in a DES model; however, agents are social and interact with others, they live in an environment, and their next actions are based on the current state of the environment. An ABS model has three elements: agents, which have attributes (static or dynamic levels, e.g. variables) and behaviors (conditional or unconditional actions, e.g., methods); interactions, which define relationships between agents; and environment, which is an external factor that affects agents and interactions (Stainsby et al. 2009; Taboada et al. 2011; Gunal 2012).

Wang (2009) used DES to evaluate ED performance under various settings of the triage process and radiology procedure process. Rahmat et al. (2013) evaluated the effects of re-triage strategy on patient waiting time based on DES modeling. The result indicated that re-triage of patients with deteriorating clinical conditions can significantly reduce their waiting time in EDs. They also suggested that additional resources such as physicians, triage officers, and supporting equipment are necessary to improve ED performance. Friesen and McLeod (2014) reviewed studies related to the investigation of patient flow and the dynamics of infection spread within hospitals using ABS and pointed out that developing robust validation and verification techniques and generating accurate models of agent behaviors and interactions are still difficult issues to solve. Kaushal et al. (2015) developed an ABS model to evaluate the effects of fast-track treatment strategies on ED performance. The results indicated that the strategy can reduce waiting patient numbers considerably.

1.5.3 SYSTEM DYNAMICS MODELING (SDM)

SDM is a popular method for modeling continuous systems and was founded by Forrester (1958). It works based on a set of differential equations that tracks instantaneous changes in a dynamic system. A typical dynamic system can be characterized by interdependence, mutual interaction, information feedback, nonlinearity, and circular causality concepts. SDM is known to be a method for strategic-level thinking since it looks at systems from higher levels to capture the whole system. It is for this reason that in SDM, we examine cohorts but not individuals. Most of the SD models in the healthcare domain are either used for persuasion purposes or for providing a framework for the evaluation of tactical studies (Homer & Hirsch 2006).

Lubyansky (2005) created an SD model to investigate how healthcare systems can best respond to surge events. The model described how hospitals and home care treatment providers can cooperate with each other to serve the communities better. Cassettari et al. (2013) built an SD model to analyze the effects of admission from ED on other hospital units and determine the critical threshold. Chong et al. (2015) applied SDM to model the patient flow in EDs in Hong Kong and try to find ways (e.g., adjust admission volumes and staff numbers) for improving ED performance. Hirsch (2004) designed an SD model of healthcare systems, which is scalable for different-sized communities with different levels of resources available. The effects of major incidents (e.g., tornadoes, explosions, and epidemics) on infrastructure systems and healthcare systems can be involved in that model. Arboleda et al. (2007) applied an SD model to represent the operation of a healthcare facility during a disaster event. The interactions between EDs, intensive care units, wards, and operating rooms are considered in the SD model. Stock-related and flow-related strategies were adopted to improve the flow of patients and reduce waiting time. Arboleda et al. (2009) integrated the network analysis of external infrastructure systems (i.e., water, power, and transportation system) and SDM of the internal capabilities of a healthcare facility. The unsatisfied demand in the infrastructure systems after disasters and the effects on hospital performance were analyzed.

Khanmohammadi et al. (2018) proposed an SD model to simulate the post-earthquake recovery process of a hospital. The seismic damage of hospital components and post-earthquake resource shortage, the flow of patients, and the dynamics of treatment operations are considered to determine the functionality and evaluate the seismic resilience of a hospital. Similarly, Li et al. (2020b) developed an SD model to simulate post-earthquake hospital functionality considering both the damages to the hospital and its recovery processes. The seismic resilience of the hospital can be evaluated based on simulation results. The proposed model was validated by a case study of a hospital in China.

1.5.4 FAULT TREE METHOD (FTM)

Apart from the simulation methods (e.g., DES, ABS, and SDM) mentioned above, the fault tree method (FTM) is another method that is widely used in system analysis of engineering systems including hospital systems. In FTM, the fault tree describes the causality relationship between system components. The minimal cut set method is used in FTM to calculate system failure probability. Lupoi et al. (2012) developed a fault tree of physical components (including the structural components and nonstructural components) in a hospital to evaluate post-earthquake performance of each component and the operation state of the hospital. Similarly, Youance et al. (2016) developed fault trees for heating system, air conditioning system, and life support system in hospitals to assess post-earthquake functionality of existing hospitals in Montreal. FTM and rapid seismic vulnerability assessment were combined by Miniati et al. (2014) to evaluate post-earthquake hospital performance and make seismic risk mitigation decisions. Jacques et al. (2014) assessed the post-earthquake performance of a hospital system using FTM based on collected data from reconnaissance. Hospital staff, stuff, and structure influence on hospital functionality were considered. The fault tree developed by Jacques et al. (2014) was then extended to estimate earthquake-induced functionality reduction in monomer hospital (Hassan and Mahmoud 2019) and healthcare networks in the Centerville community (Hassan and Mahmoud 2020). Li et al. (2018) proposed a state tree method to evaluate system performance, which is based on the concept of fault tree analysis and success path. The state tree method was then adopted by Wang et al. (2020) and Shang et al. (2020b) to evaluate the post-earthquake performance of the emergency department and the recovery process. Based on their state tree model, the component influence on the system functionality can be easily quantified.

1.6 OUTLINE OF THIS BOOK

This book contains three main parts and a schematic of the contents, and interrelationships of various chapters are illustrated in Figure 1.4:

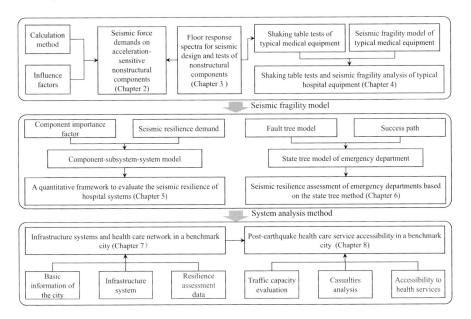

FIGURE 1.4 Outline of this book.

   I. Seismic force demands on acceleration-sensitive nonstructural components and shaking table tests of typical hospital equipment.

  II. A quantitative framework to evaluate the seismic resilience of single hospital buildings and seismic resilience assessment of emergency departments based on the state tree method.

 III. Infrastructure systems and healthcare network in a benchmark city and post-earthquake healthcare service accessibility at the network level.

Part I focuses on seismic performance and fragility of hospital nonstructural components including medical equipment. Chapter 2 describes seismic force demands on acceleration-sensitive nonstructural components, and different methods for generating floor response spectra (FRS) are summarized. Chapter 3 proposes a simplified method for the generation of FRS, and generic FRS for seismic performance tests of NSCs were established based on the analysis results. Chapter 4 conducts shaking table tests for different types of medical equipment to develop seismic fragility models, which can be used in seismic resilience assessment.

Part II of this book focuses on single hospital building-level resilience assessment. Chapter 5 proposes a quantitative framework to evaluate the seismic resilience of single hospital buildings. Chapter 6 develops a new method called the state tree method that explicitly considers the component contribution to the functionality of the emergency department to evaluate system functionality and resilience.

Part II of this book focuses on network-level resilience assessment of hospital infrastructure systems. Chapter 7 develops a Geographic Information System (GIS)–based benchmark model of a medium-sized city located in the southeastern coastal region of China, which includes the demographics, site conditions, potential hazard exposure, building inventory, and lifeline systems. Chapter 8 presents a quantitative framework that considers seismic damage to buildings and transportation networks, post-earthquake available numbers of patient beds and medical staff to evaluate post-earthquake healthcare service accessibility.

Last but not least, Chapter 9 summarizes the major research achievements and contributions reported in this monograph, based on which some innovative future research directions are proposed.

REFERENCES

Achour N, Miyajima M, Kitaura M, Price A. 2011. Earthquake-induced structural and nonstructural damage in hospitals, *Earthquake Spectra*, **27**(3), 617–634. doi:10.1193/1.3604815.

Achour N, Miyajima M, Pascale F D F, Price, A. 2014. Hospital resilience to natural hazards: Classification and performance of utilities, *Disaster Prevention and Management: An International Journal*, **23**(1), 40–52. doi:10.1108/dpm-03-2013-0057.

Achour N, Miyajima M. 2020. Post-earthquake hospital functionality evaluation: The case of Kumamoto Earthquake 2016, *Earthquake Spectra*, doi:10.1177/8755293020926180.

AminShokravi A, Heravi G. 2020. Developing the framework for evaluation of the inherent static resilience of the access to care network, *Journal of Cleaner Production*, 122123. doi:10.1016/j.jclepro.2020.122123.

Arboleda C A, Abraham D M, Lubitz R. 2007. Simulation as a tool to assess the vulnerability of the operation of a healthcare facility, *Journal of Performance of Constructed Facilities*, **21**(4), 302–312. doi:10.1061/(asce)0887-3828200721:4(302).

Arboleda C A, Abraham D M, Richard J-P P, Lubitz R. 2009. Vulnerability assessment of healthcare facilities during disaster events, *Journal of Infrastructure Systems*, **15**(3), 149–161. doi:10.1061/(asce)1076-0342200915:3(149).

Best A M, Dixon C A, Kelton W D, Lindsell C J, Ward M J. 2014. Using discrete event computer simulation to improve patient flow in a Ghanaian acute care hospital, *The American Journal of Emergency Medicine*, **32**(8), 917–922. doi:10.1016/j.ajem.2014.05.012.

Bruneau M, Chang S E, Eguchi R T, Lee G C, O'Rourke T D, Reinhorn A M, Shinozuka M, Tierney K, Wallace W A, von Winterfeldt D. 2003. A framework to quantitatively assess and enhance the seismic resilience of communities, *Earthquake Spectra*, **19**(4), 733–752. doi:10.1193/1.1623497.

Callaghan W M, Rasmussen S A, Jamieson D J, Ventura S J, Farr S. L, Sutton P D, Mathews T J, Hamilton B E, Shealy K R, Brantley D, Posner S F. 2007. Health concerns of women and infants in times of natural disasters: Lessons learned from hurricane Katrina, *Maternal and Child Health Journal*, **11**(4), 307–311. doi:10.1007/s10995-007-0177-4.

Cassettari L, Mosca R, Orfeo A, Revetria R, Rolando F, Morrison J. 2013. A system dynamics study of an emergency department impact on the management of hospital's surgery activities. In: *Proceedings of the 3rd International Conference on Simulation and Modeling Methodologies, Technologies and Applications, HA-2013*, pp. 597–604. doi:10.5220/0004617205970604.

Ceferino L, Mitrani-Reiser J, Kiremidjian A, Deierlein G, BambarénC. 2020. Effective plans for hospital system response to earthquake emergencies, *Nature Communications*, **11**, 4325. doi:10.1038/s41467-020-18072-w.

Chen H, Xie Q, Feng B, Liu J, Huang Y Chen H. 2017. Seismic performance to emergency centers, communication and hospital facilities subjected to Nepal earthquakes, 2015, *Journal of Earthquake Engineering*, 1–32. doi:10.1080/13632469.2017.1286623.

Chong M, Wang M, Lai X, Zee B, Hong F, Yeoh E, Graham C. 2015. Patient flow evaluation with system dynamic model in an emergency department: Data analytics on daily hospital records. In: *2015 IEEE International Congress on Big Data*. doi:10.1109/bigdatacongress.2015.54.

Cimellaro G P, Reinhorn A M, Bruneau M. 2010. Seismic resilience of a hospital system, *Structure and Infrastructure Engineering*, **6**(**1–2**), 127–144. doi:10.1080/15732470802663847.

Cimellaro G P, Piqué M 2014. Seismic performance of healthcare facilities using discrete event simulation models, *Geotechnical, Geological and Earthquake Engineering*, 203–215. doi:10.1007/978-3-319-06394-2_12.

Cimellaro G P, Malavisi M, Mahin S. 2016. Using discrete event simulation models to evaluate resilience of an emergency department, *Journal of Earthquake Engineering*, **21**(2), 203–226. doi:10.1080/13632469.2016.1172373.

Cimellaro G P, Malavisi M, Mahin S. 2018. Factor analysis to evaluate hospital resilience, *ASCE-ASME Journal of Risk and Uncertainty in Engineering Systems, Part A: Civil Engineering*, **4**(1), 04018002. doi:10.1061/ajrua6.0000952.

Cimellaro G P, Marasco S, Noori A Z, Mahin S A. 2019. A first order evaluation of the capacity of a healthcare network under emergency, *Earthquake Engineering and Engineering Vibration*, **18**(3), 663–677.

Dong Y, Frangopol D M. 2016. Probabilistic assessment of an interdependent healthcare-bridge network system under seismic hazard, *Structure and Infrastructure Engineering*, **13**(1), 160–170.

Duguay C, Chetouane F. 2007. Modeling and improving emergency department systems using discrete event simulation, *Simulation*, **83**(4), 311–320. doi:10.1177/0037549707083111.

Favier P, Poulos A, Vásquez J A, Aguirre P, de la Llera J C. 2018. Seismic risk assessment of an emergency department of a Chilean hospital using a patient-oriented performance model, *Earthquake Spectra*, **35**(2), 489–512. doi:10.1193/103017eqs224m.

Fawcett W, Oliveira C S.2000 Casualty treatment after earthquake disasters: Development of a regional simulation model, *Disasters*, **24**(3), 271–287.

Forrester J. 1958 Industrial dynamics: A major breakthrough for decision makers, *Harvard Business Review,* **36**(4), 37–66.

Friesen M R, McLeod R D. 2014. A survey of agent-based modeling of hospital environments, *IEEE Access*, **2**, 227–233. doi:10.1109/access.2014.2313957.

Ghanjal A, Bahadori M, Ravangard R. 2019. An overview of the health services provision in the 2017 Kermanshah earthquake, *Disaster Medicine and Public Health Preparedness*, 1–4. doi:10.1017/dmp.2018.139.

Gul M, Guneri A F. 2015. A discrete event simulation model of an emergency department network for earthquake conditions. In: 2015 6th International Conference on Modeling, Simulation, and Applied Optimization (ICMSAO). doi:10.1109/icmsao.2015.7152244.

Gunal M M. 2012. A guide for building hospital simulation models, *Health Systems*, **1**(1), 17–25. doi:10.1057/hs.2012.8.

Hassan E M, Mahmoud H. 2019. Full functionality and recovery assessment framework for a hospital subjected to a scenario earthquake event, *Engineering Structures*, **188**, 165–177. doi:10.1016/j.engstruct.2019.03.008.

Hassan E M, Mahmoud H. 2020. An integrated socio-technical approach for post-earthquake recovery of interdependent healthcare system, *Reliability Engineering & System Safety*, **106953**. doi:10.1016/j.ress.2020.106953.

Hirsch G B. 2004. Modeling the consequences of major incidents for healthcare systems. In: *Proceedings of the 22nd International Conference of the System Dynamics Society,* Oxford England, pp. 1–24.

Holling C S. 1973. Resilience and stability of ecological systems, *Annual Review of Ecology & Systematics*, **4**(1), 1–23. doi:10.1146/annurev.es.04.110173.000245.

Homer J B, Hirsch G B. 2006. System dynamics modeling for public health: Background and opportunities, *American Journal of Public Health*, **96**(3), 452–458. doi:10.2105/ajph.2005.062059.

Hoot N R, LeBlanc L J, Jones I, Levin S R, Zhou C, Gadd C S, Aronsky D. 2008. Forecasting emergency department crowding: A discrete event simulation, *Annals of Emergency Medicine*, **52**(2), 116–125. doi:10.1016/j.annemergmed.2007.12.011.

Jacques C C, McIntosh J, Giovinazzi S, Kirsch T D, Wilson T, Mitrani-Reiser J. 2014. Resilience of the Canterbury Hospital system to the 2011 Christchurch earthquake, *Earthquake Spectra*, **30**(1), 533–554. doi:10.1193/032013eqs074m.

Kadri F, Chaabane S, Tahon C. 2014. A simulation-based decision support system to prevent and predict strain situations in emergency department systems, *Simulation Modelling Practice and Theory*, **42**, 32–52. doi:10.1016/j.simpat.2013.12.004.

Kaushal A, Zhao Y, Peng Q, Strome T, Weldon E, Zhang M, Chochinov A. 2015. Evaluation of fast track strategies using agent-based simulation modeling to reduce waiting time in a hospital emergency department, *Socio-Economic Planning Sciences*, **50**, 18–31. doi:10.1016/j.seps.2015.02.002.

Khanmohammadi S, Farahmand H, Kashani H. 2018. A system dynamics approach to the seismic resilience enhancement of hospitals, *International Journal of Disaster Risk Reduction*, **31**, 220–233. doi:10.1016/j.ijdrr.2018.05.006.

Kirsch T D, Mitrani-Reiser J, Bissell R, Sauer L M, Mahoney M, Holmes W T, de la Maza F. 2010. Impact on hospital functions following the 2010 Chilean Earthquake, *Disaster Medicine and Public Health Preparedness*, **4**(2), 122–128. doi:10.1001/dmphp.4.2.122.

Komashie A, Mousavi A. 2005. Modeling emergency departments using discrete event simulation techniques. In: *Proceedings of the Winter Simulation Conference*. doi:10.1109/wsc.2005.1574570.

Li J, Wang T, Shang Q. 2018. Probability-based seismic reliability assessment method for substation systems, *Earthquake Engineering & Structural Dynamics*. doi:10.1002/eqe.3138.

Li J, Wang T, Shang Q. 2020a. Probability-based seismic resilience assessment method for substation systems, *Structure and Infrastructure Engineering.* doi:10.1080/15732479. 2020.1835998.

Li Z, Li N, Cimellaro G P, Fang D. 2020b. System dynamics modeling-based approach for assessing seismic resilience of hospitals: Methodology and a case in China, *Journal of Management in Engineering*, **36**(5), 04020050. doi:10.1061/(asce)me.1943-5479.0000814.

Lubyansky A. 2005. A system dynamics model of health care surge capacity. In: *Proceedings of the 23rd International Conference of the System Dynamics Society.* 17–21 July, 2005, Boston, USA.

Lupoi A, Cavalieri F, Franchin P. 2012. Probabilistic seismic assessment of health-care systems at regional scale. In: *15th World Conference on Earthquake Engineering* Lisboa pp. 1–10.

McGinty M D, Burke T A, Barnett D J, Smith K C, Resnick B, Rutkow L. 2016. Hospital evacuation and shelter-in-place: Who is responsible for decision-making? *Disaster Medicine and Public Health Preparedness*, **10**(3), 320–324. doi:10.1017/dmp.2016.86.

McQuade L, Merriman B, Lyford M, Nadler B, Desai S, Miller R, Mallette S. 2018. Emergency department and inpatient healthcare services utilization by the elderly population: Hurricane Sandy in the state of New Jersey, *Disaster Medicine and Public Health Preparedness*, 1–9. doi:10.1017/dmp.2018.1.

Miniati R, Capone P, Hosser D. 2014. Decision support system for rapid seismic risk mitigation of hospital systems: Comparison between models and countries, *International Journal of Disaster Risk Reduction,* **9**, 12–25. doi:10.1016/j.ijdrr.2014.03.008.

Ochi S, Tsubokura M, Kato S, Iwamoto S, Ogata S, Morita T, Hori A, Oikawa T, Kikuchi A, Watanabe Z, Kanazawa Y, Kumakawa H, Kuma Y, Kumakura T, Inomata Y, Kami M, Shineha R, Saito Y. 2016. Hospital staff shortage after the 2011 triple disaster in Fukushima Japan-An Earthquake Tsunamis and nuclear power plant accident: A case of the Soso District, *Plos ONE*, **11**(10), e0164952. doi:10.1371/journal.pone.0164952.

Paul J A, Lin L. 2012. Models for improving patient throughput and waiting at hospital emergency departments, *The Journal of Emergency Medicine*, **43**(6), 1119–1126. doi:10.1016/j.jemermed.2012.01.063.

Peek-Asa C, Kraus J F, Bourque L B, Vimalachandra D, Yu J, Abrams J. 1998. Fatal and hospitalized injuries resulting from the 1994 Northridge earthquake, *International Journal of Epidemiology*, **27**(3), 459–465. doi:10.1093/ije/27.3.459.

Peters M S. 1996. Hospitals respond to water loss during the Midwest floods of 1993: Preparedness and improvisation, *The Journal of Emergency Medicine*, **14**(3), 345–350. doi:10.1016/0736-4679(96)00031-5.

Rahmat M H, Annamalai M, Halim S A, Ahmad R. 2013. Agent-based modelling and simulation of emergency department re-triage. In: *2013 IEEE Business Engineering and Industrial Applications Colloquium (BEIAC).* doi:10.1109/beiac.2013.6560119.

Ramandi S H, Kashani. H. 2018. A framework to evaluate the resilience of hospital networks. In: *Proceedings of the Creative Construction Conference*, 30 June–3 July 2018, Ljubljana, Slovenia. doi:10.3311/CCC2018-101.

Sadegh K, Hamed K. 2017. A framework to evaluate the seismic resilience of hospitals. In: *12th International Conference on Structural Safety and Reliability*, 6–10 August, 2017, TU Wien, Vienna, Austria.

Schulze S S, Fischer E C, Hamideh S, Mahmoud H. 2020. Wildfire impacts on schools and hospitals following the 2018 California Camp Fire, *Natural Hazards.* doi:10.1007/s11069-020-04197-0.

Shahverdi B, Tariverdi M, Miller-Hooks E. 2019. Assessing hospital system resilience to disaster events involving physical damage and Demand Surge, *Socio-Economic Planning Sciences.* doi:10.1016/j.seps.2019.07.005.

Shang Q, Wang T, Li J. 2019. Seismic fragility of flexible pipeline connections in a base isolated medical building, *Earthquake Engineering and Engineering Vibration*, **18**(4), 903–916. doi:10.1007/s11803-019-0542-5.

Shang Q, Wang T, Li J 2020a. A quantitative framework to evaluate the seismic resilience of hospital systems, *Journal of Earthquake Engineering.* doi:10.1080/13632469.2020.180 2371.

Shang Q, Wang T, Li J. 2020b. Seismic resilience assessment of emergency departments based on the state tree method, *Structural Safety*, **85**, 101944. doi:10.1016/j. strusafe.2020.101944.

Shang Q, Guo X, Li Q, Xu Z, Xie L, Liu C, Li J, Wang T. 2020c. A benchmark city for seismic resilience assessment, *Earthquake Engineering and Engineering Vibration*, **19**(4), 811–826. doi:10.1007/s11803-020-0597-3.

Shirali G A, Azadian S, Saki A. 2016. A new framework for assessing hospital crisis management based on resilience engineering approach, *Work*, **54**(2), 435–444. doi:10.3233/ wor-162329.

Stainsby H, Taboada M, Luque E. 2009. Towards an agent-based simulation of hospital emergency departments. In: *2009 IEEE International Conference on Services Computing.* 21-25 September, 2009, Bangalore, India. doi:10.1109/scc.2009.53.

Swisher J R, Jacobson S H, Jun J B, Balci O. 2001. Modeling and analyzing a physician clinic environment using discrete-event (visual) simulation, *Computers & Operations Research*, **28**(2), 105–125. doi:10.1016/s0305-0548(99)00093-3.

Taboada M, Cabrera E, Iglesias M L, Epelde F, Luque E. 2011. An agent-based decision support system for hospitals emergency departments, *Procedia Computer Science*, **4**, 1870–1879. doi:10.1016/j.procs.2011.04.203.

TariVerdi M, Miller-Hooks E, Kirsch T. 2018. Strategies for improved hospital response to mass casualty incidents, *Disaster Medicine and Public Health Preparedness*, 1–13. doi:10.1017/dmp.2018.4.

Wang L. 2009. An agent-based simulation for workflow in emergency department. In: *2009 Systems and Information Engineering Design Symposium.* doi:10.1109/sieds.2009.5166148.

Wang J, Yuan B, Li Z, Wang Z. 2019a. Evaluation of public health emergency management in China: A systematic review, *International Journal of Environmental Research and Public Health*, **16**(18), 3478. doi:10.3390/ijerph16183478.

Wang T, Shang Q, Chen X, Li J. 2019b. Experiments and fragility analyses of piping systems connected by grooved fit joints with large deformability, *Frontiers in Built Environment*, **5**(49), 1–14. doi:10.3389/fbuil.2019.00049.

Wang T, Shang Q, Li J. 2020. Functionality analyses of engineering systems: One step toward seismic resilience. In: *Resilience of Critical Infrastructure Systems: Emerging Developments and Future Challenges.* CRC Press, Boca Raton, FL, pp. 163–175. doi:10.1201/9780367477394.

Wang T, Shang Q, Li J. 2021. Seismic force demands on acceleration-sensitive nonstructural components: A state-of-the-art review, *Earthquake Engineering and Engineering Vibration*, **20**(1), 39–62. doi:10.1007/s11803-021-2004-0.

WHO-PAHO. 2015. *Hospital Safety Index Guide for Evaluators* (2nd edition). World Health Organization and Pan American Health Organization, Switzerland.

Youance S, Nollet M-J, McClure G. 2016. Effect of critical sub-system failures on the post-earthquake functionality of buildings: A case study for Montréal hospitals, *Canadian Journal of Civil Engineering*, **43**(10), 929–942. doi:10.1139/cjce-2015-0428.

Zhong S, Clark M, Hou X-Y, Zang Y-L, Fitzgerald G. 2013. Development of hospital disaster resilience: Conceptual framework and potential measurement, *Emergency Medicine Journal*, **31**(11), 930–938. doi:10.1136/emermed-2012-202282.

Zhong S, Clark M, Hou X-Y, Zang Y, FitzGerald G. 2014. Validation of a framework for measuring hospital disaster resilience using factor analysis, *International Journal of Environmental Research and Public Health*, **11**(6), 6335–6353. doi:10.3390/ ijerph110606335.

2 Seismic Force Demands on Acceleration-Sensitive Nonstructural Components

2.1 RESEARCH BACKGROUND

Nonstructural components (NSCs) are parts, elements, and subsystems that are not part of the primary load-bearing system of building structures but are subject to the same seismic loading environment. Damage to NSCs may disrupt the functionality of buildings and result in significant economic losses, injuries, and casualties (Taghavi and Miranda 2003). Damage to NSCs may significantly affect the functionality of building structures. In previous earthquakes, many buildings have entirely lost their functionality not because of structural damage but because of nonstructural damage. Examples of strong earthquakes that resulted in significant damage to NSCs are the 1971 San Fernando earthquake, the 1994 Northridge earthquake, the 2010 Chile earthquake, the 2011 Tohoku Pacific earthquake, the 2013 Lushan earthquake, and the 2017 Mexico earthquake. Research on NSCs was initiated and driven by the significant effects of damage to NSCs. During the last two decades, numerous studies on NSCs have been conducted. Various seismic detailing and protection techniques have been developed and applied to NSCs to achieve multiple objectives of performance-based earthquake engineering, particularly the seismic resilience of buildings. However, the compatible performance between structural components and NSCs is difficult to achieve because of complex interfaces and the distinct mechanical properties of the two types of components.

The first step in the study of the seismic performance of NSCs is to determine the input, i.e., the floor response or the floor response spectra (FRS) at the position where the component is attached to the building. Although an NSC may be displacement and/or acceleration sensitive, most recent research has focused on the floor acceleration response spectra (abbreviated as floor response spectra, FRS herein). FRS are generated from the absolute acceleration response of a floor in a building that is excited by the input ground motion, as shown in Figure 2.1. Different from the ground acceleration spectra, FRS reflect the dynamic characteristics of the building structures. That is, the supporting structure filters out the vibrational components with frequencies different from the building's natural frequencies, whereas the vibrational components with frequencies close to the natural frequencies are amplified (Sullivan et al. 2013). Numerous studies were conducted to establish the general FRS for the seismic design of NSCs using the fundamental principles of structural dynamics. These studies demonstrated

DOI: 10.1201/9781003457459-2

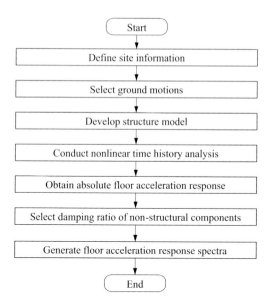

FIGURE 2.1   Development of FRS.

that the FRS were highly dependent on different parameters related to the building's characteristics and the NSC characteristics, including the location of the NSCs in the structure, the ratio of the NSC period to the building's modal periods, the damping ratio of the supporting structures and the NSCs, the structural nonlinear behavior, the interaction between the NSCs and the supporting structure, the torsional response of the supporting structures, the diaphragm flexibility of the supporting structures, the type of lateral load resisting system in the supporting building, the soil–structure interaction (SSI), and the nonlinear behavior of NSCs. The effects of several crucial influential factors on the FRS are demonstrated by the time history analyses (THA) results of an eight-story reinforced concrete (RC) frame (Wang et al. 2020). The shape and value of the FRS can be quite different for different floors. A low NSC damping ratio will result in a large FRS and vice versa. As the earthquake intensity increases, the supporting structure yields, and its floor spectral accelerations are capped by the lateral force capacity of the supporting structure, so that the FRS are reduced accordingly. A detailed discussion of these influential factors and other factors, such as the nonlinear behavior of NSCs and the soil–structure interaction, is provided subsequently.

Although several methods for generating FRS have been described in the relevant literature, none can consider all influential factors, which usually result in under- or over-estimation of the FRS. Moreover, achieving seismic resilience of buildings requires highly accurate FRS for the seismic design of the NSCs. Numerous researchers have investigated NSCs and FRS, and most of the studies are numerical analyses.

Several reviews on the seismic design of NSCs have been conducted, including the work by Chen and Soong (1988) and Villaverde (1997). NSCs are mentioned as a secondary structure or secondary systems in their review papers. There has been significant development in the design of either structure or NSCs during recent years, and many attempts have been made to generate FRS for the seismic design of NSCs. The existing state-of-the-art reviews are relatively old, and there is a need for an updated review that includes recent achievements. This chapter summarizes the progress made to date on the generation of FRS, which can be used to determine forces for the seismic design of acceleration-sensitive NSCs. Different methods for generating FRS based on single-degree-of-freedom (SDOF) models and multiple-degrees-of-freedom (MDOF) models are presented first, followed by a review of the amplification factor methods. Directly defined FRS methods and some newly developed methods are summarized next, and those included in current seismic design codes are compared. Subsequently, detailed investigations of research on the critical factors affecting FRS are outlined, including the nonlinear structural behavior, the vertical location of NSCs, and the interaction between structural components and NSCs, such as infill walls, the soil–structure interaction, the damping ratio of NSCs, and the nonlinear behavior of NSCs. The floor acceleration response and FRS obtained from experimental studies and field observations during earthquakes are discussed. In addition, the effects of vertical components of input ground motions and near-fault ground motions (NFGMs) on the FRS are presented. Finally, the major knowledge gaps to be filled are identified, and possible future research challenges are discussed. Note that most of the results presented herein were generated for linear NSCs with a 5% damping ratio, except when otherwise indicated. Moreover, the literature related to NSCs with multiple supports is not covered in this review.

2.2   DEVELOPMENT OF FLOOR RESPONSE SPECTRA

The methods to develop FRS are summarized in this section and are categorized into four groups, i.e., FRS based on SDOF models, FRS based on MDOF models, amplification factor methods, and directly defined FRS. The definitions are briefly illustrated in Figure 2.2. Most early methods fall in the first category, where either

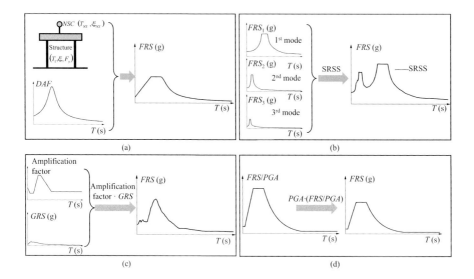

FIGURE 2.2   Definitions of the four methods to develop FRS: (a) FRS based on SDOF models, (b) FRS based on MDOF models, (c) amplification factor methods, and (d) directly defined FRS.

the supporting structure or NSCs are treated as SDOF models. Basic mechanical parameters, such as the fundamental period, the damping ratio, and the yield strength ratio, are considered. Modal superposition methods are used in methods based on MDOF models to consider the effect of different vibration modes. FRS for each mode are first developed and then combined using a modal superposition technique to obtain the final FRS. Some methods have been developed based on the ground acceleration response spectrum (GRS) or the peak ground acceleration (PGA) to facilitate the application of FRS; these methods are categorized as the third and fourth methods, respectively. Details of these methods are discussed in the following sections. The methods adopted by some seismic design codes are reviewed at the end of this section.

2.2.1 FRS BASED ON SDOF MODELS

Research on FRS generation methods began in the 1970s. Early methods usually treated the supporting structure and NSC as SDOF systems. Penzien and Chopra (1965) and Kapur and Shao (1973) were among the first to generate the FRS from the response of a supporting structure using THA. Yasui et al. (1993) derived a direct generation method, in which a smooth design FRS can be generated using the design spectra or GRS. Because of the application of the Duhamel integration, this method does not require an empirical dynamic amplification factor, which represents the spectral acceleration of the NSCs normalized by the peak floor acceleration (PFA) of the supporting structure. The obtained FRS generation formula is expressed as Equation (2.1):

$$\mathrm{FRS}\left(T_{NS},\xi_{NS}\right)=\frac{1}{\sqrt{\left[1-\left(T_S/T_{NS}\right)^2\right]^2+4\left(\xi_S+\xi_{NS}\right)^2\left(T_S/T_{NS}\right)^2}}$$
$$\sqrt{\left\{\left(T_S/T_{NS}\right)^2 S_a\left(T_S,\xi_S\right)\right\}^2+S_a\left(T_{NS},\xi_{NS}\right)^2} \quad (2.1)$$

where
  T_{NS} and ξ_{NS} are the period and damping ratio of the NSCs, respectively,
  T_S and ξ_S are the period and damping ratio of the supporting structure,
  $S_a\left(T_S,\xi_S\right)$ and $S_a\left(T_{NS},\xi_{NS}\right)$ are the values at the specific period and damping ratio in the elastic ground acceleration spectrum,
  $FRS\left(T_{NS},\xi_{NS}\right)$ are the FRS at the specific period T_{NS} and the damping ratio ξ_{NS}.

The resonance region represents the portion of the floor spectrum that includes the peak and the surrounding high spectral values. Note that in the MDOF structures, the floor spectrum usually has more than one resonance region corresponding to several natural modes. The analytical results reported by Vukobratović and Fajfar (2013) show that, outside of the resonance region, the FRS obtained by the method of Yasui et al. (1993) match very well with the FRS obtained by THA. In the resonance region, however, a substantial difference was observed. Therefore, Vukobratović and

Fajfar (2013) suggested using Equation (2.2) to calculate the FRS in the resonance region, where AMP is an empirical amplification factor in the resonance region for the considered structure, and $\dfrac{S_a(T_S, \xi_S)}{R_\mu}$ represents the value in the inelastic acceleration spectrum that can be obtained by reducing the elastic acceleration spectrum using the strength factor R_μ (Vukobratović and Fajfar 2013):

$$\text{FRS}\left(T_{NS}, \xi_{NS}\right) = \text{AMP} \cdot \frac{S_a(T_S, \xi_S)}{R_\mu} \tag{2.2}$$

Sullivan et al. (2013) used a dynamic amplification factor (DAF) to calculate the FRS in SDOF structures using Equation (2.3). The DAF is the ratio of the maximum acceleration of the NSCs to the maximum acceleration of the floor on which the NSCs are mounted. An empirical expression of the DAF was provided by Sullivan et al. (2013):

$$\text{FRS}(T_{NS}) = \begin{cases} \dfrac{T_{NS}}{T_S}\left[a_{\max}\left(\text{DAF}_{\max} - 1\right)\right] + a_{\max}, & T_{NS} < T_S \\[2ex] a_{\max}\text{DAF}_{\max}, & T_S \leq T_{NS} \leq T_e \\[1ex] a_{\max}\text{DAF}, & T_{NS} > T_e \end{cases} \tag{2.3}$$

where

$\text{FRS}(T_{NS})$ is the spectral acceleration demand for a supported component with period T_{NS}.

a_{max} is the maximum acceleration of the supporting structure (obtained for an SDOF system by dividing the structure's lateral resistance by the seismic mass).

2.2.2 FRS Based on MDOF Models

SDOF structure models cannot represent the response of multistory buildings accurately. Different methods for generating the FRS based on MDOF structural models were developed recently using a modal superposition method. Calvi and Sullivan (2014) extended the procedure of Sullivan et al. (2013) to MDOF structures as follows:

1. Determine the acceleration demand ($a_{max,m}$) on each floor for each mode m based on elastic modal analysis and the design response spectrum.
2. Apply Equation (2.3) for each vibration mode by replacing a_{max} by $a_{max,m}$ to obtain $\text{FRS}_m(T)$ for mode m.
3. For the upper floors, the FRS are calculated as the square-root-sum-of-squares (SRSS) of the modal spectra computed in Step (2).
4. For the lower floors, the FRS is the maximum between the GRS and the spectral acceleration obtained from the SRSS of the modal spectra computed in Step (2).

The above procedure distinguishes between the upper and lower floors of the building based on the hypothesis of limited higher-mode filtering that occurs for the ground motion on the lower floors of a building (Calvi and Sullivan 2014). Note that the above procedure has only been tested for elastic structures and was further improved to include the nonlinear behavior of the supporting structures by introducing modal reduction factors of the floor spectra. A modal reduction factor is defined as the ratio of the FRS from the linear response to that of the nonlinear response, which is dependent on the ductility of the supporting structure. Nonlinear regression was performed by Aragaw and Calvi (2018) to obtain the relationship between the modal reduction factor and the ductility demand of the supporting structure. Recently, the methodology proposed by Calvi and Sullivan (2014) was updated by Merino et al. (2020) to account for the nonlinear behavior of the supporting structure. They developed a code-oriented methodology, which can provide an absolute acceleration FRS and a consistent relative displacement FRS.

The method proposed by Yasui et al. (1993) can also be used for MDOF structures. For the considered number of modes, the FRS of an individual mode are obtained using Equation (2.4-1) for the off-resonance region and Equation (2.4-2) for the resonance region (Vukobratović and Fajfar 2017). In the resonance region, the FRS are defined as a product of the PFA_{ij} and AMP_i for mode i. PFA_{ij} is calculated by Equation (2.4-3) and represents a special case of Equation (2.4-1) for $T = 0$. The FRS obtained for each mode should be combined to calculate the final FRS. SRSS or CQC rules were recommended for periods between zero and the end of the resonance region of the fundamental structural mode. In the post-resonance region of the fundamental structural mode, the algebraic sum (ALGSUM) should be used.

$$\mathrm{FRS}\left(T_{NS},\xi_{NS}\right)_i = \frac{\Gamma_i\phi_{ij}}{\left|\left(T_{NS}/T_i\right)^2 - 1\right|}\sqrt{\left[\frac{S_a\left(T_i,\xi_i\right)}{R_\mu}\right]^2 + \left\{\left(T_{NS}/T_i\right)^2 S_a\left(T_{NS},\xi_{NS}\right)\right\}^2}$$

(2.4-1)

$$\left|\mathrm{FRS}\left(T_i,\xi_{NS}\right)_j\right| \leq AMP_i \cdot \left|PFA_{ij}\right|$$

(2.4-2)

$$PFA_{ij} = \Gamma_i\phi_{ij}\frac{S_a\left(T_i,\xi\right)}{R_\mu}$$

(2.4-3)

where

T_i, ξ_i, and Γ_i are the modal period, damping ration, and participation factor of mode i, respectively,

ϕ_{ij} is the mode shape for floor j of mode i,

PFA_{ij} is the peak floor acceleration on floor j for mode i,

AMP_i is the empirical amplification factor for the considered mode i, as defined in Vukobratović and Fajfar (2017).

2.2.3 AMPLIFICATION FACTOR METHODS

The amplification factor is defined as the ratio of the FRS to GRS and is usually used for generating the FRS directly from the GRS or the design response spectrum. Different amplification factor functions have been proposed in recent years. Wieser et al. (2013) developed an empirical multi-linear envelope spectral acceleration amplification function, as illustrated in Figure 2.3. The amplification function was developed based on the nonlinear time history analyses (NTHA) results of four ductile flexible special moment-resisting frames (SMRF). The function considers the influence of multiple factors, such as the period ratio of the NSCs to the first period of the supporting structure, the relative height, and the higher mode effect. The peaks located at the period ratios of 0.0, 0.3, and 1.0 correspond to the NSCs that are rigid, tuned to the second mode, and tuned to the first mode, respectively. The effective period of the structure is used to replace the elastic period (T_1) to consider the effect of structural yielding.

Five designs of RC frames based on the Indian design codes were used by Surana et al. (2018a) to calculate the FRS while considering the inelastic behavior of the frames. The authors proposed floor amplification functions (Figure 2.4) to estimate the FRS at any given height directly from the GRS if the supporting structure's first two mode shapes, the building modal periods, and the strength factor of the supporting structure are known. For short-period buildings, the amplification factor in the impact zones (i.e., the zones of the NSCs with periods between 0 and $0.5T_1$) of the second and higher modes can be considered a constant (Figure 2.4a). For long-period buildings, the peaks of the first two modes are considered (Figure 2.4b). These peaks have been represented by separate parabolic functions in the impact zone of each mode. Both elastic and inelastic structure responses were considered by Surana et al. (2018a).

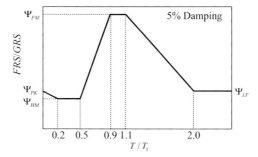

FIGURE 2.3   Spectral acceleration amplification function by Wieser et al. (2013).

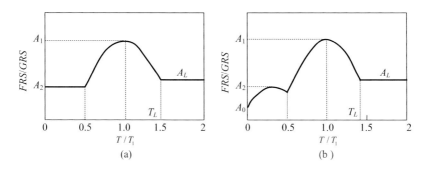

FIGURE 2.4   Spectral acceleration amplification function by Surana et al. (2018a): (a) short-period buildings and (b) long-period buildings.

2.2.4 DIRECTLY DEFINED FRS

Different from the amplification methods, directly defined FRS are based on a directly defined function of the component acceleration factor (FRS/PGA). FRS can be directly calculated using the component acceleration factor for a given PGA level. Singh et al. (2006) proposed equations for the FRS/PGA that considered the effect of possible resonance with higher modes, as shown in Figure 2.5. For a building with an unknown period, the period of the NSCs is also assumed to be unknown. A conservative estimate of the FRS/PGA is provided in Equation (2.5) for flexible NSCs with a damping ratio of 5%. For a 2% damping ratio, Equation (2.5) must be multiplied by 1.5. A detailed discussion of the effects of the NSC damping ratio is provided in Section 3.5.

$$\frac{\text{FRS}}{\text{PGA}} = 6\left(1 + 2\frac{z}{h}\right) \tag{2.5}$$

The component amplification factor (a_p) is usually used for deriving the FRS from the PFA. a_p is defined as FRS/PFA with respect to the relative height and reflects the dynamic amplification effect of a specific NSC on a specific floor. A component acceleration factor was used by Medina (2013) to quantify the maximum component acceleration demand, which is equivalent to the product of the component amplification factor (a_p) and the in-structure amplification factor (PFA/PGA). The suggested component acceleration factor for linear NSCs with a 5% damping ratio is given based on the statistical results of the NTHA. For flexible NSCs mounted near the bottom or top of the structure, a value of 3 or 12 is recommended, whereas, for rigid NSCs, a value of 2 or 4 is recommended (Medina 2013).

A simplified model for quantifying a_p was developed by Hou et al. (2018) using shaking table test results. A modification factor with a value of 1.35 was used to consider the tuning effects between the NSCs and supporting structures and prevent underestimation. However, the critical parameters used to define the simplified model were all fitted using the test results of a low-rise steel CBF building and cannot be applied to other structures.

Petrone et al. (2015) proposed a novel formulation, as shown in Figure 2.6. The parameters a, b, and a_p were determined based on the building's fundamental period T_1,

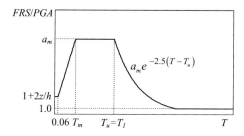

FIGURE 2.5 Component amplification factor proposed by Singh et al. (2006b).

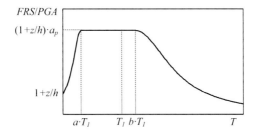

FIGURE 2.6 Shape of the floor spectra proposed by Petrone et al. (2015).

TABLE 2.1
Values of the Parameters Suggested by Petrone et al. (2015)

Fundamental Period	a	b	a_p
$T_1 < 0.5s$	0.8	1.4	5.0
$0.5s < T_1 < 1.0s$	0.3	1.2	4.0
$T_1 > 1.0s$	0.3	1.0	2.5

as shown in Table 2.1. This formulation considers the higher mode effects, and the predicted FRS are conservative for a wide range of periods, especially for periods close to T_1. The over-estimation covers the uncertainties in the calculation of structural and NSC periods. The proposed formulations are related and limited to a set of RC frames designed according to Eurocode 8 (CEN 2004). Modifications are needed when using these formulations in other structure types.

Fathali and Lizundia (2012) proposed a three-segment a_p spectrum to generate the FRS from the PFA values. The proposed spectrum was developed from recorded data in instrumented buildings and ranged from 1.0 to 2.5; it included a flat segment with a maximum value of 2.5 for medium-range periods and a nonlinear decaying segment for longer periods.

In another study, Anajafi (2018) developed expressions for the generic floor response spectrum (GFRS) based on a statistical analysis of the FRS of instrumented buildings; the expressions did not require information on the supporting building type, NSC tuning ratio, and vertical location in the building. The objective of this GFRS was to generate FRS that are neither building- nor component-specific. Anajafi (2018) suggested that the GFRS could be used for seismic tests of NSCs. Anajafi (2018) stated that in many cases when testing an NSC, no information is available on the details of a component's support and/or its attachment to the supporting building, the dynamic characteristics of the building, and/or the NSC and the location (i.e., floor level) of the attachment of the NSCs. Therefore, Anajafi (2018) concluded that it might be justifiable to advocate the use of a GFRS for testing NSCs that are neither building- nor component-specific.

2.3   CRITICAL FACTORS INFLUENCING THE FLOOR RESPONSE SPECTRA

2.3.1   EFFECTS OF NONLINEAR BEHAVIOR IN SUPPORTING STRUCTURES

The nonlinear behavior of the supporting structures has a significant influence on the floor acceleration responses. This effect has to be considered in the seismic design of NSCs in most cases, considering that the acceleration or force demands on the NSCs are usually smaller than that induced in linear supporting structures when subjected to an earthquake of the same intensity. However, FRS values may be increased sometimes, especially for NSCs, with a period between the building's modal periods. Several studies have investigated the nonlinear behavior effects of supporting structures on the acceleration demands, and the FRS used in the NSC design demonstrated that a linear structure might amplify the ground motion, whereas floor accelerations may be smaller than the ground acceleration in nonlinear structures. Taller and flexible structures with longer periods are more affected by nonlinear behavior. Lin and Mahin (1985) defined an amplification factor as the ratio of the FRS of an inelastic structure to the FRS of an elastic structure to quantify the effect of inelastic responses of the supporting structure. The amplification factor was defined at four points, i.e., A, B, C, and D, as depicted in Figure 2.7. The amplification factor values are constant before point A and after point D. Point C represents the building's fundamental period, and point B represents the maximum value. The recommended amplification factor can be used to generate the FRS considering the effects of the building's nonlinearity without conducting a nonlinear analysis.

Oropeza et al. (2010) found that the definition by Lin and Mahin (1985) may overestimate the effects of the inelastic behavior for structures with a high fundamental period and proposed an improved amplification factor based on the NTHA. Similarly, Flores et al. (2015) conducted NTHA to evaluate the FRS demand and amplification factor in steel moment frames. Zhai et al. (2016) developed a predictive model to estimate the amplification factor and quantified the nonlinearity effect of the supporting structures while considering the period of the NSCs, the period of the supporting structures, and the primary structural ductility. The predictive model of Zhai et al. (2016) was then improved by Pan et al. (2017), and a modified Park-Ang

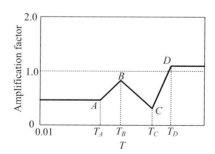

FIGURE 2.7   Amplification factors defined by Lin and Mahin (1985).

damage index was used to evaluate the damage of the supporting structures instead of the ductility factor. However, both predictive models were based on the NTHA results of specific structures and cannot be used for other types of structures.

The acceleration response modification factor (R_{acc}) was proposed by Sankaranarayanan (2007) to quantify the effects of the supporting structure's non-linear behavior. The R_{acc} factor is the reciprocal of the amplification factor defined by Lin and Mahin (1985). This factor was also used to generate the FRS while considering the effects of building nonlinearity. Sankaranarayanan (2007) also pointed out that a reduction in the FRS values occurred as a result of the structure's nonlinearity when the NSC period was close to the building's modal periods, and a greater reduction was observed near the fundamental period than in the higher modal period. In addition, an increase in the FRS values was observed when the NSC period was located between two modal periods of the building. Note that linear NSCs were considered in these studies.

2.3.2 STORY AMPLIFICATION FACTOR

The story amplification factor or in-structure amplification factor, PFA/PGA, is used for the seismic design of rigid NSCs. The seismic design force of rigid NSCs can be quite different at different locations in a building. The code provisions in ASCE 7-16 (2016) use $1 + 2z/h$ to consider the height effect. However, the linear distribution of the PFA demand in the vertical direction of the building is building-independent and overly conservative, especially for tall buildings. Moreover, the nonlinear behavior of the supporting structures is not considered in this approach. Singh et al. (2006) proposed a new formula that considered the building's fundamental period. The formula is $1 + 2z/h$ for buildings with unknown periods. When the building's fundamental period T_1 is given, the peak acceleration differs for buildings of different heights and is calculated based on the number of stories in the building.

Based on the NTHA results of five 2D steel moment-resisting frames, Akhlaghi and Moghadam (2008) demonstrated that the PFA distribution depends on the behavior of the structures, the rigidity and flexibility of the buildings, and the fundamental period. A simplified distribution of the PFA was proposed, and the building's fundamental period was considered. Although this was a simplified approach, the suggested distribution showed better results for estimating the PFA distribution than the widely used profile $1 + 2z/h$.

Surana et al. (2017) found that the PFA demand was primarily dependent on the relative height, as well as the fundamental period and the strength ratio of the supporting structure. The PFA demand on the building's roof decreased with an increase in the period and the strength ratio. A model was proposed to calculate the PFA by considering the strength ratio; this model reflects the nonlinear behavior of structures. The effect of the strength ratio was greater for buildings with fundamental periods less than 2.5 s. For buildings with periods larger than 2.5 s, it was assumed that the PFA demand was constant (i.e., PFA/PGA = 1), regardless of the floor level.

Wieser et al. (2013) proposed a model that related the PFA demand to the fundamental period of the supporting structure. The proposed model for linear structures is given in Equation (2.6-1) and that for nonlinear structures is given in Equation (2.6-2):

$$\frac{PFA}{PGA} = 1 + \frac{2.5 - T_1}{T_1}\frac{z}{h} \qquad (2.6\text{-}1)$$

$$\frac{PFA}{PGA} = 1 + \frac{2.5 - T_e}{T_e}\frac{z}{h} \qquad (2.6\text{-}2)$$

$$T_e = T_1 \sqrt{\frac{\mu}{1 + \alpha(\mu - 1)}} \qquad (2.6\text{-}3)$$

Using recorded acceleration data in buildings, Fathali and Lizundia (2012) found that the PFA/PGA relationship in Equation (13.3-1) of the ASCE 7 was suitable for buildings with periods smaller than 0.5 s but was conservative for buildings with larger periods. A new PFA/PGA relationship was defined based on the statistical results. The effects of the building's fundamental period and earthquake intensity were explicitly considered. It is important to note that most of the instrumented buildings in the United States exhibit responses in the elastic range (Fathali and Lizundia 2012). Therefore, equations derived based on the response of instrumented buildings might not be directly applicable to code-based designed buildings that are expected to exhibit an inelastic response when suffering a design earthquake ground motion (Anajafi 2018).

A simplified method was developed to rapidly estimate the floor acceleration demand in building structures that respond linearly to earthquake ground motions. The floor acceleration demands were approximated using the first three vibration modes of the building. The structure model was represented by a simplified continuum model consisting of a cantilever flexural beam connected laterally to a cantilever shear beam by axially rigid links that transmitted the horizontal forces. A comparison of the floor acceleration demand obtained from the approximation method and recorded data in six instrumented high-rise buildings indicated that the proposed method produced relatively good results with a very small computational effort and required only a small amount of information on the building. However, the method is limited to structures that remain elastic or practically elastic. The U.S. National Institute of Standards and Technology (NIST) proposed a new formula to estimate the PFA/PGA values based on recorded data and numerical analysis results (NIST 2018). The method used the PFA normalized by the PGA recorded in 44 instrumented buildings, which had experienced earthquakes with PGA > 0.15 g. Average PFA/PGA values that were computed using simplified continuous models adapted from Miranda and Taghavi (2005) were also used. The period of the supporting structure and the whiplashing effect of the higher modes were considered. NIST (2018) then suggested using the response modification factor (R) to capture the effect of the building ductility on the reduction in the acceleration demand of the NSCs. Later, Anajafi (2018) used the responses of many code-based designed buildings and proposed modifications to the expressions presented in NIST (2018) for the estimation of the R factor. Anajafi (2018) showed that equations in NIST (2018) might underestimate PFA/PGA responses for mid-height floor levels.

2.3.3   INTERACTION BETWEEN STRUCTURAL COMPONENTS AND INFILL WALLS

It is widely recognized that the presence of infill walls modifies the global structural response of buildings subjected to seismic loads and also affects PFA values, as well as the shape and maximum spectral accelerations in all stories. The effect of structural nonlinearity on the FRS is much more pronounced when infill walls are considered. Mollaioli et al. (2011) considered four-, six-, and eight-story RC frames to estimate the changes in the PFA demand at different heights of different RC frames with or without infill walls. The results indicated that the influence of the infill walls decreased as the number of stories increased. Blasi et al. (2018) observed a noticeable amplification of the FRS due to the effects of infill walls; it was also found that the infill walls generally reduced the irregularity effect in elevation, resulting in a more uniform distribution of the PFA at different heights. Surana et al. (2018b) modified their previously proposed floor spectral amplification functions (Surana et al. 2018a) for uniformly infilled and open ground story RC frames. The presence of infill walls and their mechanical properties should be considered when evaluating the seismic demand on NSCs. A set of mid-rise bare and uniformly infilled RC frame buildings was analyzed by Surana (2019) under different earthquake intensities to determine the FRS. The results verified that the effects of the infill walls were significant and should be considered, whereas the effect has not been considered in design codes. Although some studies have been conducted to investigate the effects of infill walls on the FRS, more studies are required to obtain more accurate FRS.

2.3.4   INTERACTION BETWEEN NSCS AND THE SUPPORTING STRUCTURES

For NSCs with mass ratios no larger than 1%, the dynamic interaction effect is relatively small (less than 10%) and can be neglected (Taghavi and Miranda 2008). For NSCs with a larger mass, the interaction effect may be very large for NSCs tuned to the supporting structure. However, in many cases, the supporting structure and the NSCs are decoupled and analyzed individually (i.e., decoupling analysis). Decoupling analysis sometimes provides conservative results, especially when the natural period of an NSC is close to that of the supporting structure (Chen and Soong 1988). Several studies have demonstrated that the interaction between the supporting structure and the NSCs may have a significant influence on the FRS (Taghavi and Miranda 2008). The effect of the dynamic interaction is smaller for NSCs tuned to higher modes than NSCs tuned to the fundamental mode.

  The NSC and the supporting structure were considered as a combined system by Sackman and Kelly (1979) to analyze the NSC response. Similarly, Igusa and Kiureghian (1985) presented a new method for generating the FRS in the frequency domain. The method was derived from the fundamental principles of structural dynamics, random vibrations, and the perturbation theory. Therefore, structural nonlinearity and NSCs with a large mass cannot be considered. Asfura and Kiureghian (1986) developed a cross-oscillator cross-floor response spectrum (CCFS) method to consider the interaction effect to produce more realistic design criteria for NSCs. Suarez and Singh (1987) presented a mode synthesis-based direct approach to calculate the seismic response of NSCs. Unlike the perturbation technique, the modal synthesis approach does not assume small variations. Hence, it can be used for light and heavy NSCs, regardless of the mass. Similar conclusions, as mentioned above, were drawn from these studies.

2.3.5 Damping Ratio of NSCs

The viscous damping ratios of NSCs were found to range from 1% to 30% (Aragaw and Calvi 2018). Similar to the effects on the ground motion acceleration response spectra, a low NSC damping ratio will result in large FRS, especially for NSCs with a period that is similar to that of the supporting structure, and vice versa. The effect of NSC damping on the seismic demand of NSCs needs to be properly understood. The calculated FRS should account for the most likely damping level. However, most FRS studies considered a 5% NSC damping ratio. Only a few studies have investigated the effects of NSC damping on the FRS. Aragaw and Calvi (2018) pointed out that the effects of NSC damping on the FRS can be neglected if the NSC periods are very small or very large relative to the supporting structure. Anajafi (2018) conducted a comprehensive study to determine the effects of NSC damping on the FRS. A damping modification factor (DMF) was used, which is defined as the FRS with a given damping ratio relative to the FRS for a 5% damping ratio. Numerical analyses were conducted on structures following design codes to investigate the influence of the DMF. The results indicated that NSC damping ratio and tuning period ratio (period ratio between the NSC period and the building's fundamental period) significantly affected the DMF values. Based on the numerical results, empirical equations were proposed to predict the DMF given the NSC damping ratio and tuning period ratio. Kazantzi et al. (2020) proposed a probabilistic model that incorporated the mean and lognormal standard deviation of the DMF using recorded floor acceleration data.

2.3.6 Nonlinear Behavior Effects of NSCs

Component response modification (reduction) factors are usually used in seismic design codes of NSCs to account for nonlinear behavior effects of NSCs. For instance, ASCE 7-16 (2016) uses a component response modification factor (R_p) that ranges from 1.0 to 12.0 to account for the characteristics of an NSC (including viscous damping, nonlinear behavior, and inherent over-strength). A similar behavior factor of the NSCs (q_a) was adopted in the Eurocode 8 (CEN 2004). However, the values of R_p or q_a for different NSCs were established based on engineering expertise rather than experimental tests or numerical analysis. A few studies have investigated the nonlinear behavior effects of NSCs on FRS. Igusa (1990) derived an analytical solution for the response of a 2-DOF primary-secondary system with small nonlinearities using random vibration analysis and equivalent linearization techniques. Adam and Fotiu (2000) analyzed the response of SDOF oscillators attached to a four-story frame building. The results indicated an influence on the primary response in a large frequency range when an inelastic SDOF oscillator with a low yield level was considered. Strength reduction factors were used by Villaverde (2006) to account for the nonlinear behavior of NSCs and their supporting structure. Chaudhuri and Villaverde (2008) evaluated the seismic responses of inelastic NSCs supported by moment-resisting frames. Vukobratović and Fajfar (2017) suggested the consideration of the inelastic behavior of NSCs by increasing the damping of the NSCs when generating the FRS for ductile NSCs. Filiatrault et al. (2018) used the concept of the nonstructural equivalent damping ratio to account for the nonlinear behavior of NSCs (i.e., suspended piping systems in their study). The equivalent damping ratio was derived from pseudo-static cyclic test. They found that nonlinear behavior of NSCs reduced their displacement demand. Obando and Lopez-Garcia (2018) found that NSC inelasticity significantly decreased the displacement demands on tuned NSCs, especially those tuned to the fundamental mode. Kazantzi et al. (2018) achieved similar results and found that the inelastic behavior of NSCs reduced their force and displacement demands by constant-ductility floor spectra. Anajafi (2018) generated the nonlinear FRS of several code-designed buildings considering different levels of NSC inelasticity. The results indicated that the NSC inelasticity significantly reduced the peak values of the FRS for elastic buildings in the vicinity of the building's modal periods and significantly de-emphasized the effects of the tuning period ratio, damping ratio, and the characteristics of the supporting building and ground excitation. Although previous studies have provided valuable insight into the nonlinear behavior effects of NSCs on FRS, further research is still required to incorporate this information in the seismic design of NSCs.

2.3.7 OTHER PARAMETERS THAT INFLUENCE THE ACCELERATION DEMANDS ON NSCs

It is well known that the SSI can affect the seismic response of a building and influence the FRS. However, only a few studies investigated the effects of the SSI on the floor acceleration demands and the FRS. Kennedy et al. (1981) evaluated the FRS in an MDOF pressurized water reactor auxiliary building and considered the SSI effects. Chaudhuri and Gupta (2003) included the SSI by using the sub-structure approach for determining the FRS in a 15-story shear building. The results indicated that the SSI should not be neglected when generating the FRS unless the soil is quite stiff relative to the supporting structure. Raychowdhury and Ray-Chaudhuri (2015) found that a nonlinear SSI reduced the FRS values in a steel frame. It is important to note that the NSCs were modeled as linear SDOF systems, and the dynamic interaction between the supporting structure and the NSCs was not considered in this study.

Torsional response and diaphragm flexibility of supporting structures have been investigated by a few researchers; these factors have a significant influence on the shape and peak values of the FRS according to Anajafi (2018). Qu et al. (2014) suggested that the torsional response can amplify the acceleration demands on NSCs located around the edge of the floor plan. Anajafi (2018) achieved the same results using recorded acceleration data. Qu et al. (2014) and Anajafi (2018) found that in-plane diaphragm flexibility can produce larger floor acceleration around the middle of a floor plan. These studies showed that typical buildings (even regular buildings with well-adopted floor systems) exhibited torsional behavior and diaphragm flexibility that may increase the force demands on NSCs by factors up to 2.0. Kollerathu and Menon (2017) conducted a THA of elastic and inelastic masonry structures. The results indicated that the PFA response increased with increasing diaphragm flexibility.

2.4   EFFECTS OF INPUT GROUND MOTION

2.4.1   VERTICAL GROUND MOTIONS

It is commonly assumed that buildings are flexible in the horizontal direction and rigid in the vertical direction; as a result, the vertical acceleration demands of NSCs are usually not considered. In other cases, the vertical design spectra are determined based on the horizontal spectra. A scaling factor of 0.7 or 2/3 is usually applied to the horizontal FRS to generate the vertical FRS. However, recent research on the vertical FRS showed that ignoring the effect of the vertical component of the ground motion may underestimate the vertical demand on the NSCs. Qu et al. (2014) investigated the distributions of the vertical and horizontal seismic acceleration demands on NSCs using data from instrumented buildings during earthquakes. It was found that the vertical peak floor acceleration (PFAv) was not constant in the vertical direction, which is unlike the constant distribution assumed in the ASCE 7-16 (2016). Moschen et al. (2016) statistically assessed the PFAv demands on column lines of elastic multistory steel frames. The results indicated that the vertical ground acceleration was amplified along the column line. Gremer et al. (2018) found that the PFAv amplification was significantly larger than the horizontal peak floor acceleration. The location of the beam nodes has a significant influence on the PFAv values. Therefore, it is suggested to use more than one DOF per story to evaluate the PFAv. The largest amplification in a floor occurs at the mid-span of the beam, whereas the smallest amplification occurs at the interior column. The same results were also found by Francis et al. (2017), who reported that the amplification factor might be larger than six. Guzman et al. (2017) found that the slab accelerations were generally amplified by a factor of 2.5–6.5 relative to the vertical ground acceleration, and a slab amplification factor of 4 or 5 was suggested for the vertical direction. The PFAv amplification factor ranged from 3 to 6 in a shaking table test of a full-scale steel moment frame (Ryan et al. 2016). Current design codes for NSCs neglect the amplification and are thus non-conservative.

2.4.2   NEAR-FAULT GROUND MOTIONS

Several researchers have studied the effects of NFGMs on NSCs, beginning with the study of Kennedy et al. (1981). Sankaranarayanan (2007) statistically analyzed the acceleration demands of NSCs mounted on moment-resisting frames under the excitation of NFGMs. Kanee et al. (2013) analyzed the FRS of a nonlinear frame using 49 NFGM records. Alonso-Rodríguez and Miranda (2015) investigated the floor acceleration demands in buildings subjected to NFGMs using the simplified building model developed by Miranda and Taghavi (2005). Acceleration demands were found to be sensitive to the high-frequency component of the NFGMs. The pulse duration is the most critical parameter and influences the floor acceleration demand because it induces large variations in the PFA. Zhai et al. (2016) and Pan et al. (2017) compared the FRS for 81 NFGMs and 573 ordinary ground motions. The results showed that NFGMs significantly increased the FRS of NSCs that were tuned to the supporting structure.

Most current studies on the influence of input ground motions on the FRS have primarily focused on vertical ground motions and near-fault earthquakes. More research is required on the FRS of building structures exposed to long-period earthquakes, long-duration earthquakes, and mainshock-aftershock earthquake sequences.

2.5   EXPERIMENTAL STUDIES AND FIELD OBSERVATIONS

Lepage et al. (2012) investigated the effects of floor acceleration amplification using shaking table test data of 30 multistory RC structures. The results indicated that the amplification effects were more pronounced in the upper floors and decreased with increasing earthquake intensity. Astroza et al. (2015) found that the story amplification factors of the roof of a full-scale five-story RC building were lower than the value of 3.0 suggested by ASCE 7-10. The component amplification factor a_p significantly exceeded the 2.5 limit value in ASCE 7-10 for periods near the vibration period of the supporting building. Accurate FRS of the supporting building are critical to achieving consistent and high-quality test results in shaking table tests of NSCs. Different control methods have been proposed for deriving the FRS, including the work by Maddaloni et al. (2011) and Zhou et al. (2019). Floor acceleration responses are usually reproduced based on these control methods and a loading frame. Experimental tests have been conducted on different types of NSCs, including suspended ceilings, piping systems, and other components.

Strong motion seismographs installed in structures have recorded earthquake data, which provide important references for experiments and the calculation of FRS. Naeim et al. (1998) compared the FRS calculated using data from six instrumented buildings in the 1994 Northridge earthquake with the calculated seismic demands of NSCs based on codes such as UBC-97, NEHRP-97, and FEMA-273. The results indicated that the acceleration demands on the roof exceeded the values recommended by these codes and guidelines. Acceleration data from eleven instrumented buildings in Taiwan were compared to modern code provisions by Assi et al. (2005). It was demonstrated that the code provisions $(1 + 2z/h)$ might provide under- or overestimation on the horizontal PFA of different buildings. Similar results of horizontal PFA estimations were reported by Qu et al. (2014). Wang et al. (2014) suggested that the code provisions $(1 + 2z/h)$ overestimated the PFA demands, especially when the building was in a strongly nonlinear state. Floor acceleration data measured in seven instrumented buildings were used by Lepage et al. (2012) to investigate the effects of floor acceleration amplification. Similar results to those obtained from shaking table tests were found. A similar evaluation of the design codes was conducted by Fathali and Lizundia (2012) using data recorded in 151 fixed-base buildings. Acceleration data from 59 buildings were used by Anajafi (2018) to evaluate the ASCE 7 equations for the seismic design of acceleration-sensitive NSCs. It was found that, unlike the approach in ASCE 7-16 (2016), the component amplification factor, a_p, is a function of the ratio of the NSC period to the modal periods of the supporting building, the ground motion intensity level (i.e., building inelastic behavior), and the relative height of the point of attachment of NSC to the supporting building. It was also shown that the expression provided by ASCE 7-16 for the story amplification factor ($PFA/PGA = 1 + 2z/h$) in many cases overestimates the floor acceleration responses of inelastic structures. Wang et al. (2014) and Kazantzi et al. (2018) also indicated that the a_p values in ASCE 7-16 (2016) do not necessarily provide an adequate estimation of the NSC seismic demand using recorded floor motion data.

2.6 POSSIBLE FUTURE RESEARCH CHALLENGES OF FLOOR RESPONSE SPECTRA

Past earthquake reconnaissance shows that damage to NSCs may disrupt the functionality of buildings and result in significant economic losses, injuries, and casualties. It is crucial to clarify the mechanical behavior, the damage development, and the resulting response of NSCs to determine the seismic performance of NSCs and develop appropriate designs. As a first step, many attempts have been made to investigate the NSC inputs, specifically, the FRS, which are often used to determine seismic force demands on NSCs. It is well known that the two most recent reviews on FRS were conducted about 20 years ago by Chen and Soong (1988) and Villaverde (1997). Since then, many theories and methods have been developed, and updates to seismic design codes were made. Therefore, it is necessary to summarize the progress made in the past 20 years and identify major knowledge gaps for better understanding, research, and application of FRS.

Many critical factors influencing FRS have been investigated, such as the location of the NSCs, the damping ratio of the NSCs, the interaction between NSC and structural components, the nonlinear behavior of the supporting structures and NSCs, the SSI effects, and the floor flexibility in the vertical direction and in-plane direction. The effects of input ground motions (especially vertical ground motions and near-fault earthquakes) on the FRS were also extensively investigated. Note that these factors are usually not independent. For example, the dynamics of a structure are usually affected significantly by the damping, most of which is provided by the nonlinear behavior of structures. Methods that consider either the structural and nonstructural responses do not provide an accurate prediction of the FRS. However, a combined model that considers both is challenging to solve efficiently due to the convergence problem. Therefore, although a general and more accurate FRS generation method that incorporates these influence factors will facilitate the seismic design of NSCs, more research is required to develop such a method.

Although numerous procedures have been proposed over the last few years, current building codes still do not reflect this knowledge and have not yet incorporated these procedures. One of the reasons might be the uncertainties involved in the generation methods of the FRS, which will lead to large discrepancies in the FRS and PFA demands. Some researchers have noticed this issue and tried to solve the problem. Lucchini et al. (2017) proposed a probabilistic seismic demand model (PSDM) to determine inter-story drifts and the FRS. Based on the PSDM, a method for computing uniform hazard floor acceleration spectra for linear NSCs attached to linear buildings was proposed. Epistemic uncertainties in the modal properties of the supporting structure and aleatory uncertainties caused by input ground motions were subsequently investigated by Lucchini et al. (2017). Uncertainties caused by material properties, building geometry, and live loads were considered by Perrone et al. (2020) to evaluate the FRS. However, there is currently no single method that considers all or even most uncertainties. Therefore, it remains challenging for researchers to identify the uncertainties associated with determining the FRS. Potential uncertainties include the type of ground motions, the SSI effects, the material properties of the building, NSC deployment, and modeling-related uncertainties. The aleatory and epistemic uncertainties should be considered with different exceedance probabilities in generating FRS so that the seismic design of NSCs can be incorporated into the current performance-based design procedures of buildings.

In addition, the FRS is structure-dependent, unlike the design acceleration spectra used for structures. That is, the FRS of buildings with different seismic-resisting systems, e.g., long-span structures, passively controlled structures, base-isolated structures, and super high-rise buildings, have significantly different characteristics. Pavlou and Constantinou (2006) found that the implementation of linear and nonlinear viscous dampers significantly reduced the FRS. Chalarca et al. (2019) confirmed that in most cases, the results of Pavlou and Constantinou (2006) are appropriate for viscous damping structures. However, they also found that the acceleration demand varied when the damping ratio and the velocity coefficient of the viscous dampers changed. The acceleration demand can exceed that of pure frames in some cases. Anajafi (2018) also suggested that the FRS strongly depend on the type of seismic-resisting systems. Methods for determining the FRS need to consider these differences caused by the type of seismic-resisting systems. More importantly, potential methods should also be able to discretize the degree of importance of the influential factors mentioned above for different types of buildings.

Another major research gap exists in the type of FRS. To date, most FRS have been referred to as floor acceleration response spectra. However, many acceleration-sensitive NSCs, for instance, suspended piping systems, were damaged by excessive displacement relative to the point of support on the structure. The floor acceleration response spectra solely do not provide sufficient input for all NSCs. Moreover, the floor acceleration response spectra inherently reflect the intensity and frequency of the components used as input into an NSC but cannot predict the input energy demand. It has already been observed in recent earthquakes that NSC damage was caused by long-duration shaking at relatively small intensities. It is gratifying that some studies have already started to generate relative displacement FRS or displacement demand used in NSC design. Much future work is required to conduct comprehensive studies on absolute acceleration, relative velocity, and relative displacement FRS.

Finally, FRS research based on data obtained during earthquakes is still limited. One reason is that the number of seismographs installed in building structures is still small, and another is a lack of different types of instrumented structures. However, floor response data obtained from instrumented buildings in real earthquakes will be highly useful for comparing the generated FRS with the recorded FRS and calibrating the generation method (a few sample studies were conducted by Fathali and Lizundia 2012; Anajafi 2018). Recently, the China Earthquake Administration has initiated a major project called the "China Seismic Experimental Site," in which a large number of buildings with distinct dynamic characteristics and situated in seismic-prone areas will be instrumented. The recorded strong ground motions are believed to provide insights into the FRS.

REFERENCES

Adam C, Fotiu P A. 2000. Dynamic analysis of inelastic primary-secondary systems, *Engineering Structures*, **22**(1), 58–71.

Akhlaghi H, Moghadam A S. 2008. Height-wise distribution of peak horizontal floor acceleration PHFA. In: *Proceeding of 14th World Conference on Earthquake Engineering*, October 12–17, Beijing, China.

Alonso-Rodríguez A, Miranda E. 2015. Assessment of building behavior under near-fault pulse-like ground motions through simplified models, *Soil Dynamics and Earthquake Engineering*, **79**, 47–58.

Anajafi H. 2018. Improved seismic design of non-structural components .NSCs. and development of innovative control approaches to enhance the seismic performance of buildings and NSCs, Ph.D. Dissertation. University of New Hampshire, Durham, NH.

Aragaw L F, Calvi P M. 2018. Floor response spectra in hybrid base-rocking wall buildings. In: *11th National Conference in Earthquake Engineering, Earthquake Engineering Research Institute*, Los Angeles, CA.

ASCE 7-16. 2016. *Minimum Design Loads for Buildings and Other Structures*. American Society of Civil Engineers, Reston, VA.

Asfura A, Kiureghian A D. 1986. Floor response spectrum method for seismic analysis of multiply supported secondary systems, *Earthquake Engineering & Structural Dynamics*, **14**(2), 245–265.

Assi R, McClure G, Yao G C. 2005. Floor acceleration demands for 11 instrumented buildings in Taiwan during the 1999 Chi Chi earthquake. In: *Structures Congress*, ASCE, New York, NY, pp. 1–11.

Astroza R, Pantoli E, Selva F, Restrepo J I, Hutchinson T C, Conte J P. 2015. Experimental evaluation of the seismic response of a rooftop-mounted cooling tower, *Earthquake Spectra*, **31**(3), 1567–1589.

Blasi G, Perrone D, Aiello M A. 2018. Fragility functions and floor spectra of RC masonry infilled frames: Influence of mechanical properties of masonry infills, *Bulletin of Earthquake Engineering*, **16**(12), 6105–6130.

Calvi P M, Sullivan T J. 2014. Improved estimation of floor spectra in RC wall buildings. In: *10th National Conference in Earthquake Engineering*, Earthquake Engineering Research Institute, Anchorage, AK.

CEN. 2004, *Eurocode 8: Design of Structures for Earthquake Resistance-Part 1: General Rules, Seismic Actions and Rules for Buildings*, EN 1998-1. CEN, Brussels, Belgium.

Chalarca B, Perrone D, Filiatrault A. 2019., Floor acceleration demand on steel moment resisting frame buildings retrofitted with linear and nonlinear viscous dampers. In: *4th International Workshop on the Seismic Performance of Non-structural Elements, SPONSE*, Pavia, Italy, 22–23 May.

Chaudhuri S R, Gupta V K. 2003. Mode acceleration approach for generation of floor spectra including soil-structure interaction, *Journal of Earthquake Technology*, **40**(213), 99–115.

Chaudhuri S R, Villaverde R. 2008. Effect of building nonlinearity on seismic response of non-structural components: A parametric study, *Journal of Structural Engineering*, **134**(4), 661–670.

Chen Y, Soong T T. 1988. State-of-the-art review seismic response of secondary systems, *Engineering Structures*, **10**(4), 218–228.

Fathali S, Lizundia B. 2012. Evaluation of ASCE/SEI 7 equations for seismic design of non-structural components using CSMIP records. In: SMIP12 Seminar on Utilization of Strong-Motion Data, Sacramento, CA.

Filiatrault A, Perrone D, Merino R J, Calvi G M. 2018. Performance-based seismic design of non-structural building elements, *Journal of Earthquake Engineering*. doi:10.1080/136 32469.2018.1512910.

Flores F X, Lopez-Garcia D, Charney F A. 2015a. Acceleration demands on nonstructural components in special steel moment frames. In: 11th Congress of Earthquake and Seismic Engineering in Chile, Santiago de Chile, 18–20 March.

Francis T C, Hendry B C, Sullivan T J. 2017. Vertical spectral demands on building elements induced by earthquake excitation. In: New Zealand Society for Earthquake Engineering Annual Conference, Wellington, New Zealand.

Gremer N, Moschen L, Adam C, Medina R A. 2018. Horizontal and vertical acceleration demand on moment-resisting steel frames. In: 16th European Conference on Earthquake Engineering, June 18–21, Thessaloniki, Greece.

Guzman Pujols J C, Ryan K L. 2017. Slab vibration and horizontal-vertical coupling in the seismic response of low-rise irregular base-isolated and conventional buildings, *Journal of Earthquake Engineering*, **24**, 1–36.

Hou H T, Fu W Q, Wang W, Qu B, Chen Y Y, Chen Y S, Qiu C X. 2018. Horizontal seismic force demands on nonstructural components in low-rise steel building frames with tension-only braces, *Engineering Structures*, **168**, 852–864.

Igusa T. 1990. Response characteristics of inelastic 2-DOF primary-secondary system, *Journal of Engineering Mechanics*, **116**(5), 1160–1174.

Igusa T, Kiureghian A D. 1985. Generation of floor response spectra including oscillator-structure interaction, *Earthquake Engineering & Structural Dynamics*, **13**(5), 661–676.

Kanee A R T, Kani I M Z, Noorzad A. 2013. Elastic floor response spectra of nonlinear frame structures subjected to forward-directivity pulses of near-fault records, *Earthquakes and Structures*, **5**(1), 49–65.

Kapur K K, Shao L C. 1973. Generation of seismic floor response spectra for equipment design. In: ASCE Specialty Conference on Structural Design of Nuclear Power Plant Facilities, Chicago, IL, pp. 29–71.

Kazantzi A, Vamvatsikos D, Miranda E. 2018. Effect of yielding on the seismic demands of nonstructural elements. In: 16th European Conference on Earthquake Engineering, June 18–21, Thessaloniki, Greece.

Kazantzi A K, Vamvatsikos D, Miranda E. 2020. The effect of damping on floor spectral accelerations as inferred from instrumented buildings, *Bulletin of Earthquake Engineering*, **18**, 2149–2164.

Kennedy R P, Short S A, Newmark N M. 1981. The response of a nuclear power plant to near-field moderate magnitude earthquakes. In: Transactions of the 6th International Conference on Structural Mechanics in Reactor Technology, Palais des Congres, North-Holland Pub. Co. for the Commission of the European Communities, Paris, France.

Kollerathu J A, Menon A. 2017. Role of diaphragm flexibility modelling in seismic analysis of existing masonry structures, *Structures*, **11**, 22–39.

Lepage A, Shoemaker J M, Memari A M. 2012. Accelerations of nonstructural components during nonlinear seismic response of multistory structures, *Journal of Architectural Engineering*, **18**(4), 285–297.

Lin J, Mahin S A. 1985. Seismic response of light subsystems on inelastic structures, *Journal of Structural Engineering*, **111**(2), 400–417.

Lucchini A, Franchin P, Mollaioli F. 2017. Median floor acceleration spectra of linear structures with uncertain properties, *Earthquake Engineering & Structural Dynamics*, **46**(12), 2055–2060.

Maddaloni G, Ryu K P and Reinhorn A M. 2011. Simulation of floor response spectra in shake table experiments, *Earthquake Engineering & Structural Dynamics*, **40**(6), 591–604.

Medina R A. 2013. Seismic design horizontal accelerations for non-structural components. In: Vienna Congress on Recent Advances in Earthquake Engineering and Structural Dynamics, August 28–30, Vienna, Austria.

Merino R J, Perrone D, Filiatrault A. 2020. Consistent floor response spectra for performance-based design of nonstructural elements, *Earthquake Engineering & Structural Dynamics*, **49**(3), 261–284.

Miranda E, Taghavi S. 2005. Approximate floor acceleration demands in multistory buildings. I: Formulation, *Journal of Structural Engineering*, **131**(2), 203–211.

Mollaioli F, Lucchini A, Bazzurro P, Bruno S, De Sortis A. 2011. Floor horizontal acceleration demand on reinforced concrete frames. In: Fib Symposium Prague 2011—Concrete Engineering for Excellence and *Efficiency*.

Moschen L, Medina R A, Adam C. 2016. Vertical acceleration demands on column lines of steel moment-resisting frames, *Earthquake Engineering & Structural Dynamics*, **45**(12), 2039–2060.

Naeim F, Lobo R, Martin J A. 1998. Performance of nonstructural components during the January 17, 1994 Northridge earthquake-case studies of six instrumented multi-story buildings. In: Seminar on Seismic Design, Retrofit, and Performance of Nonstructural Components ATC-29-1, San Francisco, CA, pp. 107–119.

NIST 2018, Recommendations for Improved Seismic Performance of Nonstructural Components, NIST GCR 18-917-43. Applied Technology Council, Redwood City, CA.

Obando J C, Lopez-Garcia D. 2018. Inelastic displacement ratios for nonstructural components subjected to floor accelerations, *Journal of Earthquake Engineering*, **22**(4), 569–594.

Oropeza M, Favez P, Lestuzzi P. 2010. Seismic response of nonstructural components in case of nonlinear structures based on floor response spectra method, *Bulletin of Earthquake Engineering*, **8**(2), 387–400.

Pan X, Zheng Z, Wang Z. 2017. Estimation of floor response spectra using modified modal pushover analysis, *Soil Dynamics and Earthquake Engineering*, **92**, 472–487.

Pavlou E, Constantinou M C. 2006. Response of nonstructural components in structures with damping systems, *Journal of Structural Engineering*, **132**(7), 1108–1117.

Penzien J, Chopra A K. 1965. Earthquake response of an appendage in multi-storey building. In: Third World Conference of Earthquake Engineering, New Zealand.

Perrone D, Brunesi E, Filiatrault A, Nascimbene R. 2020. Probabilistic estimation of floor response spectra in masonry infilled reinforced concrete building portfolio, *Engineering Structures*, **202**, 109842.

Petrone C, Magliulo G, Manfredi G. 2015. Seismic demand on light acceleration-sensitive nonstructural components in European reinforced concrete buildings, *Earthquake Engineering & Structural Dynamics*, **44**(8), 1203–1217.

Qu B, Goel R K, Chadwell C B. 2014. Evaluation of ASCE/SEI 7 provisions for determination of seismic demands on nonstructural components. In: 10th National Conference in Earthquake Engineering, Earthquake Engineering Research Institute, Anchorage, AK.

Raychowdhury P, Ray-Chaudhuri S. 2015. Seismic response of nonstructural components supported by a 4-story SMRF: effect of nonlinear soil-structure interaction, *Structures*, **3**, 200–210.

Ryan K L, Soroushian S, Maragakis E, Sato E, Sasaki T, Okazaki T. 2016. Seismic simulation of an integrated ceiling-partition wall-piping system at E-defense. I: Three-dimensional structural response and base isolation, *Journal of Structural Engineering*, **142**(2), 04015130.

Sackman J L, Kelly J M. 1979. Seismic analysis of internal equipment and components in structures, *Engineering Structures*, **1**(4), 179–190.

Sankaranarayanan R. 2007. Seismic response of acceleration-sensitive nonstructural components mounted on moment resisting frame structures, Ph.D. Dissertation. University of Maryland, College Park.

Singh M P, Moreschi L M, Suárez L E, Matheu E E. 2006. Seismic design forces II: Flexible nonstructural components, *Journal of Structural Engineering*, **132**(10), 1533–1542.

Suarez L E, Singh M P. 1987. Floor response spectra with structure-equipment interaction effects by a mode synthesis approach, *Earthquake Engineering & Structural Dynamics*, **15**(2), 141–158.

Sullivan T J, Calvi P M, Welch D P. 2013. Estimating roof-level acceleration spectra for single storey buildings. In: 4th ECCOMAS Thematic Conference on Computational Methods in Structural Dynamics and Earthquake Engineering, Kos Island, Greece.

Surana M, Singh Y, Lang D H. 2017. Effect of response reduction factor on peak floor acceleration demand in mid-rise RC buildings, *Journal of the Institution of Engineers. India), Series A*, **98**(**1–2**), 53–65.

Surana M, Singh Y, Lang D H. 2018a. Floor spectra of inelastic RC frame buildings considering ground motion characteristics, *Journal of Earthquake Engineering*, **22**(3), 488–519.

Surana M, Pisode M, Singh Y, Lang D H. 2018b. Effect of URM infills on inelastic floor response of RC frame buildings, *Engineering Structures*, **175**, 861–878.

Surana M. 2019. Evaluation of seismic design provisions for acceleration-sensitive non-structural components, *Earthquakes and Structures*, **16**(5), 611–623.

Taghavi S, Miranda E. 2003. Response Assessment of Nonstructural Building Elements, PEER Report 2003/05. Pacific Earthquake Engineering Research Center, Berkeley, CA.

Taghavi S, Miranda E. 2008. Effect of interaction between primary and secondary systems on floor response spectra. In: 14th World Conference on Earthquake Engineering, October 12–17, Beijing, China.

Villaverde R. 1997. Seismic design of secondary structures: State of the art, *Journal of Structural Engineering*, **123**(8), 1011–1019.

Villaverde R. 2006. Simple method to estimate the seismic nonlinear response of nonstructural components in buildings, *Engineering Structures*, **28**(8), 1209–1221.

Vukobratović V, Fajfar P. 2013. A method for direct generation of floor acceleration spectra for inelastic structures. In: Transactions of the 22nd International Conference on Structural Mechanics in Reactor Technology, SMiRT 22, Paper no 215, San Francisco, CA, 18–23 August.

Vukobratović V, Fajfar P. 2017. Code-oriented floor acceleration spectra for building structures, *Bulletin of Earthquake Engineering*, **15**(7), 3013–3026.

Wang X, Astroza R, Hutchinson T, Bachman R. 2014. Seismic demands on acceleration-sensitive nonstructural components using recorded building response data-case study. In: Tenth US National Conference on Earthquake, Earthquake Engineering Research Institute, Anchorage, AK, 2014.

Wang T, Shang Q X, Li J C. 2020. Case study of floor acceleration response spectra in reinforced concrete frames using different methods. In: 17th World Conference on Earthquake Engineering, September 13–18, Sendai, Japan.

Wieser J, Pekcan G, Zaghi A E, Itani A, Maragakis M. 2013. Floor accelerations in yielding special moment resisting frame structures, *Earthquake Spectra*, **29**(3), 987–1002.

Yasui Y, Yoshihara J, Takeda T, Miyamoto. 1993. Direct generation method for floor response spectra. In: Transactions of the 12th international conference on structural mechanics in reactor technology, SMiRT 12, 15–20 August 1993, Stuttgart, Germany, Paper no **K13/4**, pp. 367–372.

Zhai C H, Zheng Z, Li S, Pan XL, Xie L L. 2016. Seismic response of nonstructural components considering the near-fault pulse-like ground motions, *Earthquakes and Structures*, **10**(5), 1213–1232.

Zhou H M, Shao X Y, Tian Y P, Xu G X, Shang Q X, Li H Y, Wang T. 2019. Reproducing response spectra in shaking table tests of nonstructural components, *Soil Dynamics and Earthquake Engineering*, **127**, 105835.

3 Floor Response Spectra for Seismic Design and Tests of Nonstructural Components

3.1 RESEARCH BACKGROUND

Nonstructural components (NSCs) are those systems and components attached to the floors, roof, and walls of a building or industrial facility. They are not part of a building's main structural load-bearing system, but are nonetheless subjected to the same dynamic environment during an earthquake. Researchers have considered the seismic performance of NSCs via experimental tests and numerical modeling. This study is motivated by the extensive nonstructural damage recorded after recent earthquakes. Direct and indirect economic loss caused by the damage of NSCs can be greater than that caused by structural components (SCs). It should also be noted that NSCs are easily damaged when suffering earthquake intensities that are not strong enough to induce damage to SCs. The functionality of buildings, especially strategic buildings, such as hospitals, will be compromised by damage to NSCs, including equipment. Moreover, it is widely recognized that damaged NCSs, such as falling tiles from suspended ceilings, may also increase the risk to the occupants' safety. Depictions of nonstructural damage caused by recent earthquakes are provided in Figure 3.1.

The seismic evaluation and design of NSCs are necessary to protect these components from damage and preserve the post-earthquake functionality of buildings. To this end, the first step in the study of the seismic performance of an NSC is to determine the seismic input, i.e., the floor response time history or floor response spectra, at the position where the NSC is attached to the supporting structure. Typically, the floor acceleration response time history or the floor acceleration response spectra (FRS) can be used for the seismic design of acceleration-sensitive NSCs, the inter-story displacement response time history or the floor displacement response spectra can be used for displacement-sensitive NSCs, and the relative velocity response time history or the floor velocity response spectra can be used for velocity-sensitive NSCs. An FRS is the acceleration response spectra obtained from the absolute acceleration time history of a typical floor in a supporting structure. Different from the ground acceleration spectra, the FRS reflects the dynamic characteristics of building structures, i.e., the supporting structure will filter out the vibrational components with frequencies different from the natural vibration frequencies of the supporting structure, and will amplify those close to the natural frequencies (Sullivan et al. 2013). D'Angela et al. (2021) evaluated the seismic damage of unanchored NSCs subjected

 DOI: 10.1201/9781003457459-3

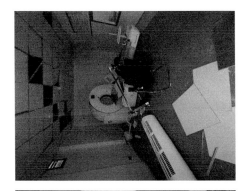

FIGURE 3.1 Nonstructural damage caused by recent earthquakes.

to ground motion and floor acceleration response motion. The fragility assessment results indicated that, in most cases, the amplification and filtering effects of the supporting structures will increase the damage probability of unanchored NSCs. Over the years, the acceleration demands of NSCs supported by single-story structures and multi-story structures have been investigated based on the fundamental principles of structural dynamics. A recently detailed literature review of FRS can be found in the publication of Wang et al. (2021); they pointed out that FRS are highly dependent on several structural and nonstructural parameters, including the location of the NSC in the structure, the ratio of the component period to the structural modal periods, the higher mode effect, the damping ratio of the supporting structure and NSCs, etc. Although several methods for the generation of FRS have been reported in the relevant literature, none can thoroughly consider all these influencing factors, which usually results in the under- or over-estimation of the FRS. Moreover, the existing methods were usually developed for European and American structures, while those for structures designed based on Chinese codes are limited. Moreover, the seismic design concepts for structures differ between Chinese codes and European and American codes. This again motivated this study, in which the floor acceleration responses in Chinese buildings are analyzed, and a method for the generation of FRS used for the seismic design of acceleration-sensitive NSCs is developed.

In this study, a simplified method based on the modification of the methodology developed by Vukobratović and Fajfar (2017) is proposed for the generation of FRS. The component dynamic amplification factors (DAFs) that reflect the amplification effects of NSCs on the floor acceleration responses play an important role in the generation of FRS. Therefore, three reinforced concrete (RC) moment-resisting frames with different heights are designed to evaluate the component DAFs. An empirical fitted formula of the DAFs is then proposed to generate more accurate FRS for structures designed according to Chinese seismic design codes (GB 50011-2010, 2010). The height factors, floor response spectra, peak component acceleration, and nonstructural damping effects are also evaluated using the results calculated by time history analyses (THAs). The results are then used to validate the proposed DAFs and demonstrate the feasibility and effectiveness of the proposed method.

3.2 PROTOTYPE RC MOMENT-RESISTING FRAMES

3.2.1 STRUCTURE DESIGN AND NUMERICAL MODELING

Three RC moment-resisting frame models of 4-, 8-, and 12-story structures (strong column-weak beam frames) were designed in accordance with the relevant Chinese codes to evaluate the floor acceleration responses. All prototype frames had the same layout, as shown in Figure 3.2. The height of the bottom story was 4.1 m, that of the second story was 3.8 m, and that of all other stories was 3.6 m, as depicted in Figure 3.3. The dimensions of the column and beam cross-sections and more detailed information on the prototype frames can be found in Qu et al. (2019).

Most of the studies on FRS use the lumped mass-spring model (LMSM) as a simplified model for the seismic response analysis of moment-resisting frames. However, the over- or underestimation of the seismic response of a moment frame may occur

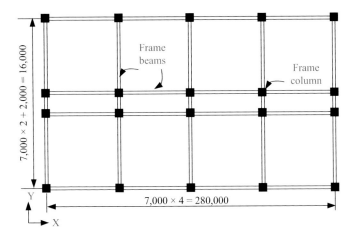

FIGURE 3.2   The layout of the prototype RC frames (unit: mm).

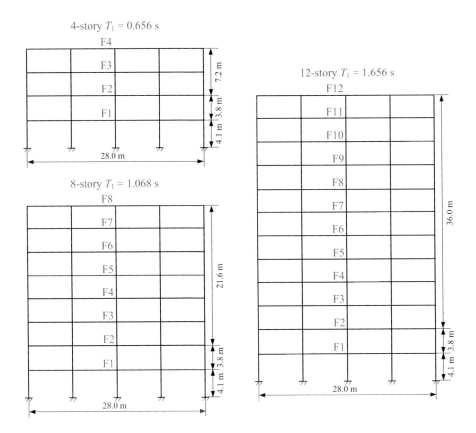

FIGURE 3.3   The lateral view of the considered RC frames.

when the LMSM is used, as this model cannot explicitly consider the column-to-beam strength ratio and the stiffness ratio (Qu et al. 2019). For instance, a larger column-to-beam stiffness ratio leads to a more uniform distribution of inter-story drifts. Models that fail to correctly consider the column-to-beam strength ratio and the stiffness ratio (e.g., the LMSM) may over- or underestimate the potential of deformation and damage concentration in a moment frame (Paulay and Priestley 1992). Moreover, the adoption of a three-dimensional (3D) detailed finite element model (DFEM) to conduct dynamic THAs is time-consuming. As a compromise between the DFEM and LMSM, the fishbone model (Masayoshi et al. 2010) is adopted in this study to calculate the floor acceleration responses of the supporting structures. Numerical fishbone models of the prototype frames were developed on the OpenSees platform. The accuracies of the fishbone models of the three frames were evaluated via nonlinear static analysis, dynamic analysis, and incremental dynamic analysis [detailed information is available in Qu et al. (2019)]. The first three modal periods and modal masses of the numerical models are listed in Table 3.1, and the modal shapes are shown in Figure 3.4, where z is the height of the structure with respect to the base, h is the height of the roof of the structure with respect to the base, and z/h represents the relative height. The mass participation coefficients of the first modes of the structures decrease as the number of stories increases, but are all larger than 80% (except for the value of 79.63% for the 12-story frame), thereby implying that the first mode is the dominant mode for the considered structures.

TABLE 3.1
The Modal Periods and Modal Masses of the Numerical Models

Structure	Mode	Period (s)	Modal Mass (tons)	Mass Participation Coefficient	
4-story	1	0.656	490.4	88.16%	99.67%
	2	0.206	50.1	9.02%	
	3	0.123	13.9	2.49%	
8-story	1	1.068	1010.2	83.60%	96.05%
	2	0.355	112.1	9.28%	
	3	0.202	38.3	3.17%	
12-story	1	1.656	1458.5	79.63%	98.16%
	2	0.574	294.4	16.07%	
	3	0.340	45.1	2.46%	

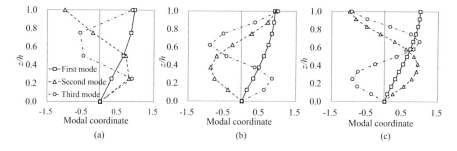

FIGURE 3.4   The modal shapes for the (a) 4-, (b) 8-, and (c) 12-story frames.

3.2.2   GROUND MOTION RECORDS AND TIME HISTORY ANALYSES

THAs were carried out using 20 spectra-compatible ground motions to obtain the floor acceleration responses of the considered structures. The motion set was composed of 15 recorded ground motions selected from the PEER-NGA database and five artificial ground motions. The recorded ground motion dataset consists of motions from earthquakes ranging in magnitude from 5.38 to 7.90. Detailed information is summarized in Table 3.2. The artificial ground motions were generated following the design spectra used in China. Figure 3.5 presents the acceleration response spectra of the selected ground motions along with the mean response spectra and design response spectra scaled to the service-level earthquake (SLE) intensity with a 63.2%

TABLE 3.2
Information about the Recorded Ground Motions Used for the THAs

Motion	Earthquake Name	Date	M_w	D (km)	R (km)	PGA (cm/s²)
GM6	RSN5275_CHUETSU_NIGH01EW	20070716	6.80	9.0	0.05	115.85
GM7	RSN5259_CHUETSU_NIG013EW	20070716	6.80	9.0	0.05	105.81
GM8	RSN4457_MONTENE.GRO_ULA000	19790415	7.10	7.0	0.25	179.61
GM9	RSN3494_CHICHI.06_TCU108N	19990925	6.30	16.0	0.03	64.27
GM10	RSN3275_CHICHI.06_CHY036N	19990925	6.30	16.0	0.04	135.80
GM11	RSN2951_CHICHI.05_CHY039N	19990922	6.20	10.0	0.05	57.65
GM12	RSN2937_CHICHI.05_CHY015N	19990922	6.20	10.0	0.05	61.88
GM13	RSN2115_DENALI_PS11-66	20021103	7.90	8.9	0.13	70.48
GM14	RSN1762_HECTOR_ABY090	19991016	7.13	14.8	0.08	178.28
GM15	RSN1528_CHICHI_TCU101-E	19990920	7.62	8.0	0.05	207.68
GM16	RSN1338_CHICHI_ILA050-E	19990920	7.62	8.0	0.05	63.55
GM17	RSN1148_KOCAELI_ARE090	19990817	7.51	16.0	0.087	131.51
GM18	RSN1115_KOBE_SKI090	19950116	6.90	17.9	0.125	124.16
GM19	RSN1048_NORTHR_STC090	19940117	6.69	17.5	0.163	334.65
GM20	RSN392_COALINGA_B-CHP000	19830611	5.38	2.4	0.15	56.28

M_w, moment magnitude; D, focal depth; R, epicentral distance; PGA, peak ground acceleration.

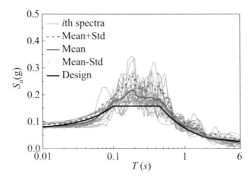

FIGURE 3.5   The response spectra of the selected ground motion records.

exceedance probability in 50 years. The mean spectra for this set of ground motions are comparable in shape to the design spectra for a Type-III site in an Intensity-VIII region in China.

3.3 THE TIME HISTORY ANALYSIS RESULTS FOR ELASTIC STRUCTURES

3.3.1 HEIGHT FACTOR

The peak floor acceleration (PFA) is the FRS of a very stiff NSC at 0.0 s, and the ratio of the PFA to the peak ground acceleration (PGA), namely, PFA/PGA, is defined as the height factor for a given floor. The current code formulas usually implicitly simplify the variation of the PFA over the building height as a linear relationship. The definitions of PFA/PGA in different seismic design codes, including ASCE 7 (ASCE 7-16 2016), Eurocode 8 (CEN 2004), and GB 50011-2010 (2010), are, respectively, given by Equations (3.1-1)–(3.1-3). To better understand the height effects and quantify the distribution of PFA values over the building height, the height factors of the three frames were calculated. The mean PFA/PGA values along the height of the supporting structures calculated from 20 THAs are depicted in Figure 3.6a, while the individual results using different ground motion inputs are depicted in Figure 3.6b–d, in which "Std" indicates the standard deviation. The variation of the mean PFA/PGA values along the

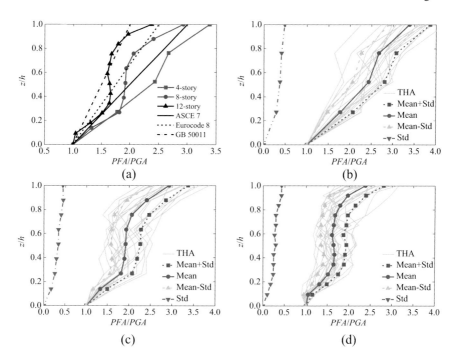

FIGURE 3.6 The height factors determined by the THAs: (a) comparison between the three frames and current codes; (b) PFA/PGA in the 4-story frame; (c) PFA/PGA in the 8-story frame; and (d) PFA/PGA in the 12-story frame.

height of the frame is obviously nonlinear. The PFA/PGA values from ASCE 7 almost envelope the numerical results, excluding those of the 4-story frame, while those from GB 50011-2010 (2010) may significantly underestimate the PFA demands along the building height. The PFA/PGA values in structures with longer vibration periods (i.e., the 12-story frame) were found to be generally smaller than those in structures with shorter periods (i.e., the 4-story frame). For longer-period structures, the PFA/PGA values provided in the current provisions are more conservative. A significant dispersion of the PFA/PGA values calculated using different ground motion inputs was observed in the three frames, as depicted in Figure 3.6b–d. In general, the linear assumption in the current codes may lead to excessive over- or under-evaluation in some cases.

$$\frac{PFA}{PGA} = 1 + 2\frac{z}{h} \tag{3.1-1}$$

$$\frac{PFA}{PGA} = 1 + 1.5\frac{z}{h} \tag{3.1-2}$$

$$\frac{PFA}{PGA} = 1 + \frac{z}{h} \tag{3.1-3}$$

3.3.2 FLOOR RESPONSE SPECTRA

The NSCs considered here are those that can be represented by single-degree-of-freedom (SDOF) systems with masses that are significantly smaller than the total mass of the supporting structure (usually the ratio of the mass of the NSC to the total mass of the supporting structure is less than 1%). The NSCs are assumed to be elastic SDOF systems with a single point of attachment. It is worth noting that the dynamic interaction effects between light NSCs and the supporting structure can be neglected (Medina et al. 2006). The decoupled procedure for generating FRS via THAs is presented in Figure 3.7. Ground motions compatible with the design acceleration spectra are selected and used as input for the THA. For a given structure and ground motion,

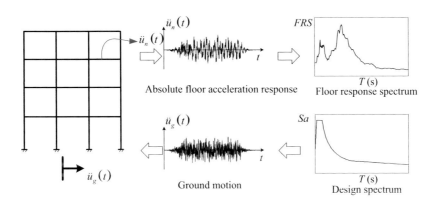

FIGURE 3.7 The decoupled procedure for the generation of the FRS.

the absolute acceleration response at the selected floor levels is obtained and used as the input for an SDOF oscillator to develop its corresponding FRS.

The mean FRS results of the 20 ground motions corresponding to a 5% nonstructural damping ratio (ξ_{NS}) normalized by the PGA are exhibited in Figure 3.8a–c. The accelerations on each floor of the 4-story frame; the second, fourth, sixth, and eighth floors of the 8-story frame; and the third, sixth, ninth, and twelfth floors of the 12-story frame are used to provide a near-complete picture of the distribution of the FRS over the height of the supporting structure. Figure 3.8 reveals that the FRS/PGA ratio varied with the vibration period of the NSCs, which is neglected

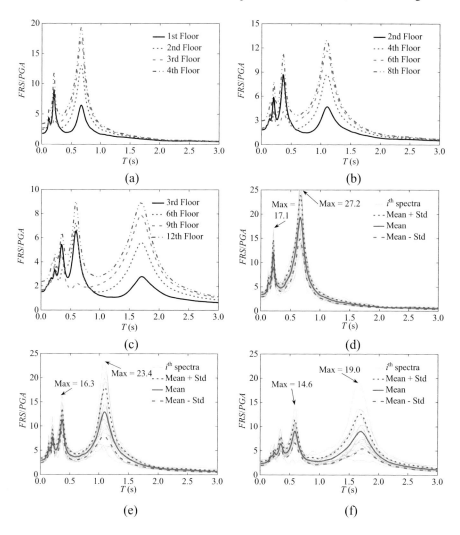

FIGURE 3.8 The FRS calculated by the THAs: (a) 4-story frame; (b) 8-story frame; (c) 12-story frame; (d) FRS of the top floor in the 4-story frame; (e) FRS of the top floor in the 8-story frame; and (f) FRS of the top floor in the 12-story frame.

in the current ASCE 7 and GB 50011-2010 codes. The peak FRS/PGA values occurred when the NSC was in tune with one of the modal periods (T_1, T_2, or T_3) of the supporting structure, and the resonance effects with higher modes of the supporting structure can almost be neglected. This is representative of the behavior observed along the height of all the structures, which highlights the importance of considering the vibration period of the i-th mode in the quantification of the FRS. It was also found that the FRS values, especially the peak FRS values, are greatly dependent on floor levels of the NSCs; the FRS values were generally found to be larger for higher floor levels. As the floor level increased, the FRS values of the fundamental period of the supporting structures increased rapidly as compared with those of higher mode periods and the period regions away from the modal periods. The gray lines in Figure 3.8d–f represent individual FRS of the top floor for each of the 20 ground motions. The ASCE 7 limit of 4.0 for FRS/PGA was found to be significantly exceeded (see values such as 17.1 and 27.2), and this phenomenon has also been observed in instrumented buildings, as reported in a previous study (Anajafi 2018). Furthermore, a significant dispersion about the FRS was also identified for the results of different ground motions.

3.3.3 Peak Component Acceleration

The maximum FRS over the period range of interest is defined as the peak component acceleration (PCA). The PCA normalized to the PGA (PCA/PGA), assuming an elastic NSC ($R_p = 1.0$) with an importance factor of $I_p = 1.0$, can be defined as Equation (3.2) according to the definition in ASCE 7. The lower and upper limits of the PCA/PGA values are 0.75 and 4.0, respectively. If the upper limit is not applied, the maximum PCA/PGA ratio predicted by Equation (3.2-1) is 7.5.

$$\frac{PCA}{PGA} = \frac{F_p}{0.4 S_{DS} W_p} = a_p \left(1 + 2\frac{z}{h} \right) \tag{3.2-1}$$

$$0.75 \leq \frac{PCA}{PGA} \leq 4.0(7.5) \tag{3.2-2}$$

where F_p is the horizontal seismic design force applied at the NSCs; S_{DS} is the short-period, 5%-damped, pseudo-spectral acceleration for the building site; W_p is the operating weight of the NSCs; and a_p is the component DAF, which varies from 1.00 (for rigid NSCs) to 2.50 (for flexible NSCs).

Figure 3.9 exhibits the relationship between the mean values of PCA/PGA and the relative height. Moreover, the ASCE 7 equation without applying the upper limit of 4.0, which is denoted as "ASCE w/o limit," is also illustrated. The PCA responses were found to be smaller for longer-period structures, but the ASCE 7 definition was exceeded at all relative heights for the three structures, especially at the top floor. Thus, the code-defined linear (without limit) or bilinear (with limit) relationships between PCA/PGA and z/h cannot predict the peak acceleration response of NSCs distributed along the building height.

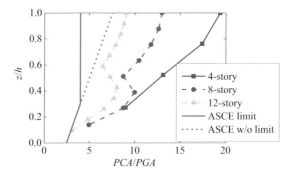

FIGURE 3.9   The comparison of the mean PCA/PGA values with the ASCE 7 definition.

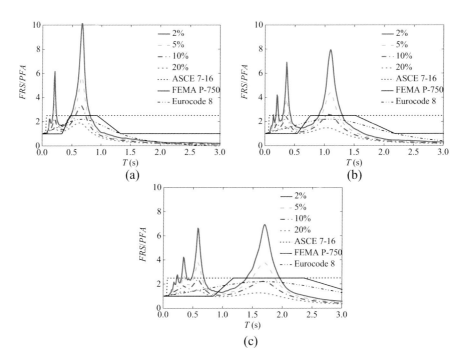

FIGURE 3.10   The top-floor component DAFs at different nonstructural damping ratios: (a) 4-story frame; (b) 8-story frame; and (c) 12-story frame.

3.3.4   Component Dynamic Amplification Factor

NSCs can be characterized by different periods of vibration and damping properties. Aragaw and Calvi (2021) found that NSCs can vary from very stiff (with periods ranging from 0.01 to 0.2 s) to very flexible (with periods ranging from 1.0 to 2.0 s), with damping ratios ranging from 1% to 30%. Thus, the effects of ξ_{NS} on the seismic demand of NSCs must be properly understood. Figure 3.10 presents the top-floor FRS normalized by the corresponding PFA considering different ξ_{NS} values (namely

2%, 5%, 10%, and 20%). This range of ξ_{NS} is appropriate for the characterization of NSCs. The FRS/PFA values represent the component DAF. The DAFs were found to be strongly dependent on the selected ξ_{NS} values. As was reasonable to expect, lower ξ_{NS} values resulted in higher amplification factor values. The influence of ξ_{NS} was found to be more pronounced at the modal periods (T_1, T_2, and T_3) of the supporting structure. For very short and very long periods, as well as the periods between modal periods, the influence of ξ_{NS} was found to be very low or even negligible. The design codes in ASCE 7 state that all flexible NSCs ($T > 0.06$ s) have a component acceleration amplification factor of 2.5, and this value is 2.0 in GB 50011-2010. This value is constant for all components with periods greater than 0.06 s. However, the code definitions significantly underestimate the DAF for periods close to the vibration periods of the supporting structures, especially for NSCs with low damping ratios (i.e., lower than 10%). For the commonly used ξ_{NS} value of 5%, the maximum DAF values for the top floors of the 4-story, 8-story, and 12-story structures are respectively 5.72, 4.42, and 3.80. This underestimation can also be found in the definitions of Eurocode 8 and FEMA P-750 (FEMA P-750 2009), as depicted in Figure 3.10.

3.4 PROPOSED METHOD FOR THE GENERATION OF FLOOR RESPONSE SPECTRA

3.4.1 OVERVIEW OF THE METHODOLOGY BY VUKOBRATOVIĆ AND FAJFAR

FRS can be quite different for the off-resonance and resonance regions, as is evident from Figures 3.8 and 3.10. A method for the generation of FRS based on that originally developed by Yasui et al. (1993) was proposed for off-resonance and resonance regions by Vukobratović and Fajfar (2015). The resonance region is the period range in which the peaks of the FRS occur and is defined by a ±15% broadening of the peak FRS period. The use of the resonance region instead of the resonance period is intended to consider the uncertainties related to the determination of the natural vibration periods of the supporting structure.

   In the case of multi-story structural systems, FRS are usually obtained via the combination of the FRS calculated for individual vibration modes, and the horizontal displacements of floors are used as the degrees of freedom. For floor j of mode i, the value of the FRS corresponding to the off-resonance region is calculated by Equation (3.3), where i represents the vibration mode and j is the degree of freedom. The feasibility of Equation (3.3) has been demonstrated in Vukobratović and Fajfar (2017) using structures designed according to Eurocode 8.

$$\text{FRS}_{ij}\left(T, \xi_{NS}\right) = \frac{\Gamma_i \phi_{ij}}{\left|\left(T / T_i\right)^2 - 1\right|} \sqrt{S\left(T_i, \xi\right)^2 + \left\{\left(T / T_i\right)^2 S\left(T, \xi_{NS}\right)\right\}^2}, \qquad (3.3)$$

where Γ_i is the modal participation factor of mode i and is defined by Equation (3.4), ϕ_{ij} is the mode shape for floor j of mode i, T and ξ_{NS} are the period and damping ratio of the NSCs, T_i and ξ are the period and damping ratio of the supporting structure, $S\left(T_i, \xi\right)$ and $S\left(T, \xi_{NS}\right)$ are the values at the specific period and damping ratio in the input elastic acceleration spectra, e.g., the design acceleration spectra, and

$FRS_{ij}\left(T,\xi_{NS}\right)$ represents the floor response spectra at the specific period T and the damping ratio ξ_{NS} corresponding to floor j of mode i.

$$\Gamma_i = \frac{\{\phi_i\}^T[M]\{1\}}{\{\phi_i\}^T[M]\{\phi_i\}}, \tag{3.4}$$

where $\{\phi_i\}$ is the mode shape of mode i, $[M]$ is the mass matrix, and $\{1\}$ is the vector in which each element is equal to unity.

In the resonance region, the FRS is defined as a product of the peak floor acceleration (PFA_{ij}) and an empirical amplification factor AMP_i for the considered mode i, as given by Equation (3.5). The amplification factor AMP_i is the maximum value of DAF mentioned previously. The results from Vukobratović and Fajfar (2015) indicate that the most important parameter influencing AMP_i is the nonstructural damping ratio ξ_{NS}, which cannot be accurately determined. Vukobratović and Fajfar (2017) proposed an empirical definition of AMP_i based on the parametric study results of European buildings, as given by Equation (3.5-3). It should be noted that ξ_{NS} is entered as a percentage when calculating AMP values using Equation (3.5-3). The FRS obtained for individual modes are then combined by using the standard square root of the sum of the squares (SRSS) or complete quadratic combination (CQC) modal combination rules.

$$FRS_{ij}\left(T,\xi_{NS}\right)=AMP_i \cdot \left|PFA_{ij}\right| \tag{3.5-1}$$

$$PFA_{ij}=\Gamma_i\phi_{ij}S\left(T_i,\xi\right) \tag{3.5-2}$$

$$AMP_i=\begin{cases} 2.5\sqrt{10/\left(5+\xi_{NS}\right)} & T_i/T_C = 0.0 \\ \text{linear between } AMP_i\left(T_i/T_C = 0.0\right) \text{ and } AMP_i\left(T_i/T_C = 0.2\right) & 0 \leq T_i/T_C \leq 0.2 \\ 10/\sqrt{\xi_{NS}} & T_i/T_C \geq 0.2 \end{cases} \tag{3.5-3}$$

where PFA_{ij} is the PFA value for floor j of mode i, AMP_i is the empirical amplification factor value of mode i, and T_C is the characteristic period of the ground motion.

Figure 3.11 exhibits the comparison of the response spectra for the top floors of the 4-, 8-, and 12-story structures calculated by the methodology developed by Vukobratović and Fajfar (2017) and those calculated by the THAs. Table 3.3 reports the relative errors of the top-floor height factors and the maximum top-floor FRS values between the results calculated by the THAs and the unmodified methodology Vukobratović and Fajfar (2017). When comparing the results of the method proposed by Vukobratović and Fajfar (2017) to the THA values of structures designed by Chinese codes, excessive underestimation is evident. This underestimation is mainly due to the differences between the amplification factors of structures designed by Chinese and European codes, as well as the effects of the modal combination rules, which will be discussed later.

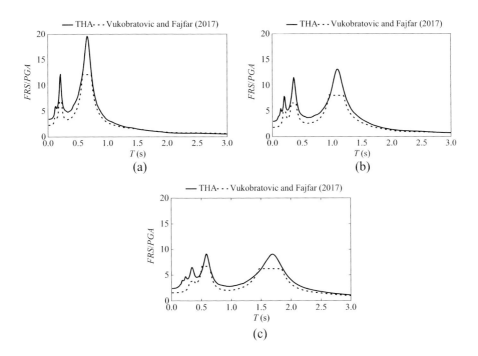

FIGURE 3.11 The FRS of the top floors of the structures: (a) 4-story frame; (b) 8-story frame; and (c) 12-story frame.

TABLE 3.3

The Relative Error between the Results Calculated by the THAs and the Unmodified Methodology

Error	PFA/PGA	Maximum FRS of the Top Floor
4-story frame	−34%	−38%
8-story frame	−39%	−38%
12-story frame	−35%	−26%

Source: Vukobratović and Fajfar (2017).

3.4.2   MODIFICATION OF THE METHOD BY VUKOBRATOVIĆ AND FAJFAR

In this work, two main modifications are made to the method developed by Vukobratović and Fajfar (2017) to achieve more accurate predictions of the floor acceleration responses of structures designed according to Chinese codes. The first modification is a new definition of AMP, and the other is related to the modal combination rule. The different definitions of AMP may result in quite different FRS. Another definition of AMP for the case of $T_i / T_C \geq 0.2$ defined by Vukobratović and Fajfar (2016) is given by Equation (3.6-1). Figure 3.12 exhibits comparisons of different definitions of AMP provided in related literature and codes (Sullivan et al. 2013; Oropeza et al. 2010; Eurocode 8; Welch 2016). The AMP results calculated

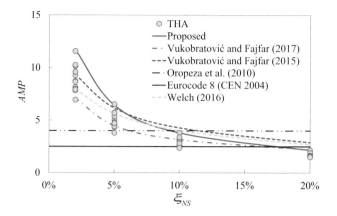

FIGURE 3.12   The definition of the amplification factor.

from the THAs of different structures and different floor levels are also provided in Figure 3.12. Moreover, the empirical fitted formula of the amplification factor (Equation (3.6-2) based on the envelope of the results calculated by the THAs is proposed and will be used instead of Equation (3.5-3) in the generation of FRS.

$$AMP_i = 18\left(1 + \xi_{NS}\right)^{-0.6} \tag{3.6-1}$$

$$AMP_i = 30.136\left(1 + \xi_{NS}\right)^{-0.864} \tag{3.6-2}$$

As mentioned previously, FRS obtained for individual modes are usually combined via the standard SRSS or CQC modal combination rules. However, several researchers have noted that these standard combination rules sometimes do not work well for the combination of absolute accelerations (Jiang et al. 2015; Pozzi and Der 2015). In this study, it is suggested that the algebraic summation (ALGSUM) of the first three modes be used to obtain more accurate floor acceleration responses, as given by Equation (3.7). It should be noted that the final value of PFA is determined as the maximum value between that determined using Equation (3.7-1) and the value of the PGA to avoid underestimation, as given by Equation (3.8).

$$PFA_j = \sum_{i=1}^{3} PFA_{ij} = \sum_{i=1}^{3} \Gamma_i \phi_{ij} S\left(T_i, \xi\right) \tag{3.7-1}$$

$$FRS_j\left(T, \xi_{NS}\right) = \sum_{i=1}^{3} FRS_{ij}\left(T, \xi_{NS}\right) = \sum_{i=1}^{3} AMP_i \cdot \left|PFA_{ij}\right| = \sum_{i=1}^{3} AMP_i \cdot \left|\Gamma_i \phi_{ij} S\left(T_i, \xi\right)\right|$$

$$\tag{3.7-2}$$

$$PFA_j = \max \begin{cases} \displaystyle\sum_{i=1}^{3} \Gamma_i \phi_{ij} S\left(T_i, \xi\right) \\ \\ PGA \end{cases} \tag{3.8}$$

3.5   VALIDATION OF THE PROPOSED METHOD

The mean results calculated by the THAs were considered as accurate results to validate the proposed method. Figure 3.13a–c presents comparisons between the height factors calculated using the proposed method, the unmodified methodology

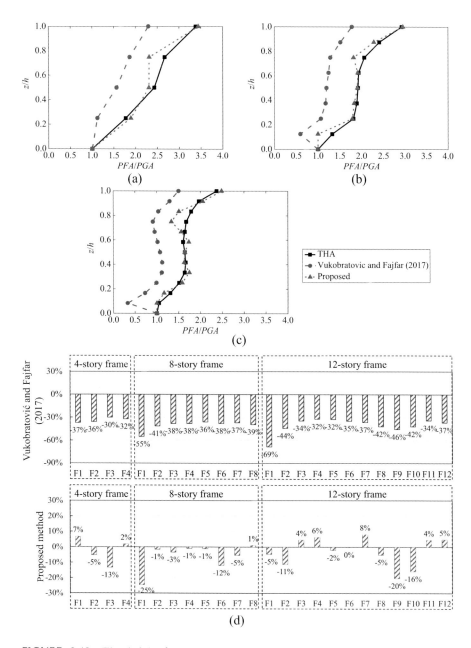

FIGURE 3.13 The height factor comparisons: (a) 4-story frame; (b) 8-story frame; (c) 12-story frame; and (d) relative error.

(Vukobratović and Fajfar 2017), and the THAs. The proposed approach was found to perform well, as the shape and magnitude of the predicted PFA matched the results calculated by the THAs. The relative errors were calculated as the ratio of the gap between the results calculated by the THAs and predicted values to the results calculated by the THA and are shown in Figure 3.13d. The maximum relative errors predicted by the proposed method for the 4-, 8-, and 12-story frames were, respectively, −13.31%, −24.6%, and −20.3%. Furthermore, the average absolute value of the relative errors was 6.8%, which is acceptable in engineering practice. The unmodified methodology (Vukobratović and Fajfar 2017) yielded smaller prediction results, and the relative errors were larger than those of the proposed method. Thus, the proposed method can predict relatively well the distribution of floor accelerations along the height of the supporting structure with low computational effort.

The FRS obtained using the proposed method were compared with those calculated by the THAs to highlight the accuracy and reliability of the proposed method. Figure 3.14 presents the FRS of the 4-, 8-, and 12-story structures for the NSCs with a damping ratio of 5%. Based on a comparison with the results predicted by the unmodified methodology, as shown in Figure 3.11, the proposed method can include the peaks related to the modal periods of the structure, and a relatively better prediction effect was achieved for the off-resonance region. The root-mean-square error [RMSE, Equation (3.9)] was employed to quantify the prediction quality of the FRS obtained by different methods. Table 3.4 reports the RMSE values for the top-floor FRS/PGA values predicted by the unmodified methodology and the proposed modified method. Both the resonance and off-resonance regions are considered here, and the relative errors at the peak points of the FRS/PGA curves are also provided. The modified method was found to achieve better prediction results than the unmodified methodology. It should be noted that NSCs with a vibration period away from the resonance region with the supporting structures are usually selected for the seismic design to avoid resonance effects. Therefore, the prediction results are acceptable from the engineering perspective, although there was a gap between the resonance region prediction results and the THA results.

$$\text{RMSE} = \sqrt{\frac{1}{N}\sum_{i=1}^{N}\left[y_{THA}(i) - y_{predicted}(i)\right]^2}, \tag{3.9}$$

where N is the total number of considered points, and $y_{THA}(i)$ and $y_{predicted}(i)$ are the FRS/PGA values determined by the THAs and predicted by different methods, respectively.

The predicted top-floor FRS for different ξ_{NS} values are presented in Figure 3.15. For NSCs with a high damping ratio (i.e., $\xi_{NS} = 20\%$), the proposed method sometimes slightly overestimated the FRS at the resonance region, and for NSCs with a low damping ratio, the proposed method sometimes slightly underestimated the FRS at the resonance region. However, in general, the proposed method was found to sufficiently consider the effects of the nonstructural damping ratio for the prediction of the FRS.

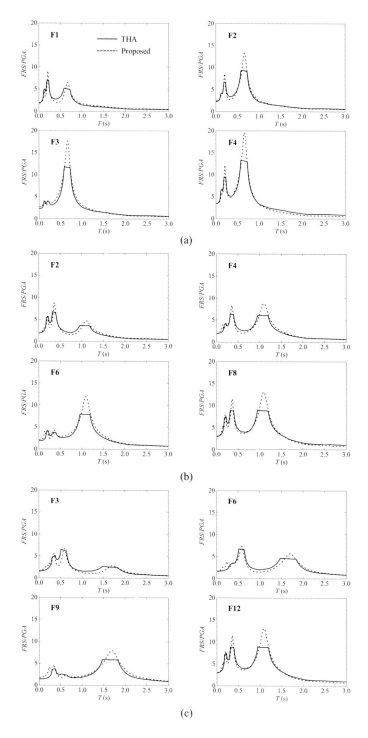

FIGURE 3.14 The FRS determined by the THAs and the proposed method: (a) 4-story frame; (b) 8-story frame; and (c) 12-story frame.

TABLE 3.4
The Relative Errors and RMSE Values

Structure	Relative Error at the Peak Point of the FRS		RMSE at the Resonance Region of T_1		RMSE at One Part of the Off-Resonance Region (for $T_2 < T < T_1$)	
	Vukobratović and Fajfar (2017)	Proposed	Vukobratović and Fajfar (2017)	Proposed	Vukobratović and Fajfar (2017)	Proposed
4-story	38%	32%	4.6976	3.8694	9.9726	2.8397
8-story	38%	32%	3.3044	2.6914	8.4801	3.1238
12-story	26%	7%	1.7490	1.3253	6.4681	4.1865

3.6 DEVELOPMENT OF GENERAL FLOOR RESPONSE SPECTRA FOR SHAKING TABLE TESTS OF NONSTRUCTURAL COMPONENTS

3.6.1 ACCELERATION DEMANDS OF NSCs DEFINED IN DIFFERENT CODES

The seismic design of NSCs was first included in the Applied Technology Council report (ATC 1978). Subsequently, more seismic design codes have proposed the use of seismic design methods for NSCs. The DAF ranges from 1.0 to 2.5 for rigid NSCs (with a period shorter than 0.06 s) and flexible NSCs (with a period larger than 0.06 s) in ASCE 7-16 (2016), whereas the value for flexible NSCs is 2.0 in GB 50011-2010 (2010). The DAF in NZS 1170.5 (2004) is related to the NSC period without considering the supporting structure period, as shown in Figure 3.16 and Table 3.5, whereas the factors in the Eurocode 8 (CEN 2004) and NEHRP (FEMA P-750 2009) are related to both the periods of the NSCs and the supporting structure. However, the available codes and guidelines, such as ASCE 7-16 (2016) and GB 50011-2010 (2010), are based on past experiences and engineering judgment instead of test results or numerical analysis.

Note that the above definitions can be used to calculate the equivalent static force for the seismic design of NSCs. The PCA demand in ASCE 7-16 (2016) is calculated as the product of DAF and PFA. Many studies have evaluated the definition of the PFA and DAF and suggested some modifications should be made to obtain a more accurate result (e.g., Anajafi 2018). Dynamic analyses, including the linear dynamic analysis procedure (Section 12.9 of ASCE 7-16, 2016), nonlinear response history procedure (Chapters 16, 17, and 18 of ASCE 7-16, 2016), FRS methods (Section 13.3.1.4.1 of ASCE 7-16, 2016), and alternate FRS methods (Section 13.3.1.4.2 of ASCE 7-16, 2016), are also permitted by ASCE 7-16 (2016) to determine the design forces for NSCs. In the FRS method, the FRS is calculated for the design earthquake at each floor level based on a seismic response history analysis for each ground motion. In the alternate FRS method, the DAF is used. The peak acceleration response for the i-th mode is calculated as the product of the modal participation factor, the spectral acceleration, and the DAF value for the i-th mode. The FRS takes the maximum value of the peak acceleration response at each modal period of the building (at least the first three modes) but is not less than the GRS values.

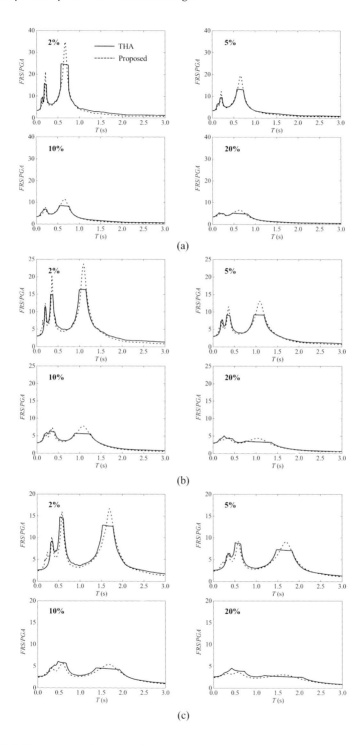

FIGURE 3.15 The top-floor FRS for different nonstructural damping ratios: (a) 4-story frame; (b) 8-story frame; and (c) 12-story frame.

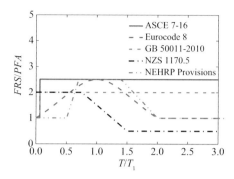

FIGURE 3.16 The component dynamic amplification factor in different design codes [$T_1 = 1.0\,$s is used in the Eurocode 8 (CEN 2004)].

TABLE 3.5

Definition of Nonstructural Dynamic Amplification Factor in Current Codes

Code	Dynamic Amplification Factor	Values
GB 50011-2010	2.0 for flexible NSCs and 1.0 for rigid NSCs	1.0 or 2.0
FEMA P-750	$T/T_1 \leq 0.5$	1.0
	$0.5 < T/T_1 \leq 0.7$	$7.5\,(T/T_1) - 2.75$
	$0.7 < T/T_1 \leq 1.4$	2.5
	$1.4 < T/T_1 \leq 2$	$-2.5\,(T/T_1) + 6.0$
	$T/T_1 > 2$	1.0
ASCE 7-16	$T \leq 0.06\,$s	1.0
	$T > 0.06\,$s	2.5
Eurocode 8	$\dfrac{\dfrac{3(1+z/h)}{1+(1-T/T_1)^2} - 0.5}{\dfrac{3(1+z/h)}{2} - 0.5}$	Related to the floor height and the first period of the supporting structures
NZS 1170.5	$T \leq 0.75\,$s	2.0
	$0.75 < T < 1.5\,$s	$2\,(1.75 - T)$
	$T \geq 1.5\,$s	0.5

Note: T is the natural period of the NSCs, and T_1 is the fundamental natural period of the supporting structure.

3.6.2 COMPONENT DYNAMIC AMPLIFICATION FACTOR BY NONLINEAR TIME HISTORY ANALYSIS

The top-floor component DAFs of 4-, 8-, and 12-story structures under different earthquake intensities (PGA =.07, 0.20, and 0.40 g) are shown in Figure 3.17. It can be found that the floor acceleration response spectra (or component dynamic

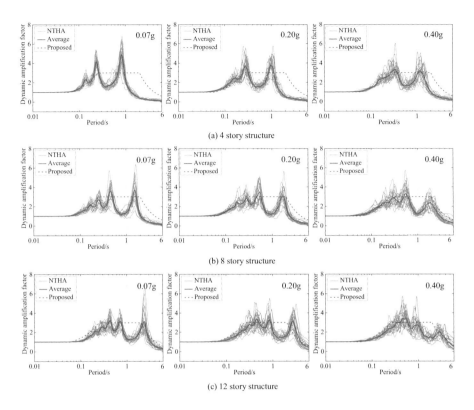

(a) 4 story structure

(b) 8 story structure

(c) 12 story structure

FIGURE 3.17 Dynamic amplification factor of nonstructural components.

amplification factor spectra) usually have multiple peak value points, and the corresponding periods of NSCs are around the natural periods of the supporting structures. These peak values of component dynamic amplification factor are resulted from the resonance effects between the supporting structure and NSCs. It can also be found that the generated FRS under the excitation of different ground motions is also different and shows some discreteness. The peak points tend to move to higher periods with an increase in the earthquake intensities, in which the supporting structures suffered nonlinear responses and the earthquake input energy was dissipated. This "frequency shift effect" was mainly due to the elongation of the natural periods of the supporting structures under the plastic deformation stage. The "frequency shift effect" was more significant for the first natural period of the supporting structures and was not so obvious for higher periods. The maximum value of component DAFs was found to not always occur around the first natural period of the supporting structures, e.g., the maximum value was found around the second period of the 12-story structure. Moreover, the resonance effects between NSCs and the third and higher modes of the supporting structure can be ignored. In summary, the definitions of DAF in existing design codes are usually related to the natural period of the NSCs, and the results in Figure 3.17 indicated that the peak values of DAF are much higher than those defined in existing codes (Table 3.5). The definitions in existing design codes are intended for seismic design instead of seismic shaking table tests, and thus, these definitions cannot be used for shaking table tests directly.

3.6.3 Definition of the Generic Floor Response Spectra

NSCs are easy to be damaged under earthquakes, and the seismic damage of NSCs will have great influence on post-earthquake normal use functionality of buildings. Shaking table tests can be used to investigate the seismic response and damage modes of NSCs. FRS are used to determine inputs of shaking table tests on acceleration-sensitive NSCs. The selection of FRS will have great influence on test results. Based on the nonlinear time history analysis of standard frame structures under different earthquake intensities, FRS for seismic performance tests of NSCs were established. Considering the fact that the seismic performance tests are usually for NSCs with little information about the supporting structure and distribution floor, the proposed response spectra are composed of short period section (0.0–0.06 s), linear increase section (0.06–0.25 s), platform section (0.25–2.0 s), and decrease section (after 2.0 s), as shown in Equation (3.10) and Figure 3.17. It should be noted that the selection of 0.06 s is based on the existing code definition about flexible and rigid NSCs in ASCE 7-16 and Gb 50011-2010. The DAF of the platform section is selected as 3.0, the same as that defined in YD 5083 (2006) and Anajafi (2018). The period range of platform section is selected from 0.25 s to 2.0 s, which covers the most widely used building structures to consider the resonance effects between the NSCs and the supporting structures. It can be found in Figures 3.17 and 3.18 that the proposed FRS can almost envelope the results from NTHA. Considering that the definition of Equation (3.10) is based on the results of 4-, 8-, and 12-story RC frame structures, the applicability for other types of structures (e.g., masonry structures and shear wall structures) still needs calibration.

$$\beta = \begin{cases} 1.0 & 0.0 \leq T \leq 0.06 \\ 10.53T+0.37 & 0.06 < T \leq 0.25 \\ 3.0 & 0.25 < T \leq 2.0 \\ 3.0\left(\dfrac{2}{T}\right)^{1.5} & T > 2.0 \end{cases} \qquad (3.10)$$

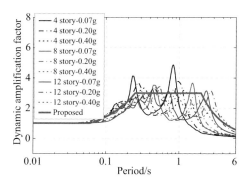

FIGURE 3.18 Median values of dynamic amplification factor of nonstructural components.

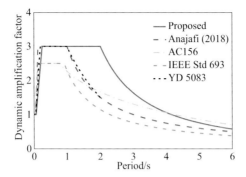

FIGURE 3.19 Comparison of dynamic amplification factors of nonstructural components.

TABLE 3.6
Dynamic Amplification Factors Defined in YD 5083 (2006)

Frequency/Hz	0.5	1.0	5.0	10.0	20.0	50.0
Dynamic amplification factors	1.5	3.0	3.0	1.5	1.0	1.0

The comparisons of the proposed DAF definition in this study and that defined in AC156, IEEE Std 693 (2005), YD 5083 (2006), and Anajafi (2018) are shown in Figure 3.19. The definition by Anajafi (2018) is given in Equation (3.11), while that defined in YD 5083 (2006) is listed in Table 3.6. In addition, YD 5083 (2006) suggested that the DAFs between the defined frequency points in Table 3.6 are linearly distributed in the logarithmic coordinate system.

$$2\beta = \begin{cases} 2.0 & 0.0 \le T \le 0.06 \\ 66.7T - 2.0 & 0.06 < T \le 0.11 \\ 6.0 & 0.11 < T \le 1.0 \\ \dfrac{6}{T} & T > 1.0 \end{cases} \qquad (3.11)$$

3.7   SUMMARY AND CONCLUSIONS

Nonstructural components (NSCs) are becoming increasingly important in preserving post-earthquake functionality to achieve seismic-resilient buildings. Floor acceleration responses are used as the seismic design input for acceleration-sensitive NSCs and therefore should be carefully determined. Although an NSC could be displacement- and/or acceleration-sensitive, most of the previous related research has focused on the floor acceleration response.

This study presented the basic characteristics of floor acceleration response spectra (FRS) using results from the THAs of three elastic RC frames designed according to Chinese codes. The NSCs considered were those that can be represented by SDOF systems with a single point of attachment. It was demonstrated that the linear assumption of the height factor values in the current codes may lead to excessive over- or underestimation. The peak FRS values generally occur when the component is in tune with the supporting structure, and the location of the component in a building also has a significant influence on the spectra values. Moreover, the peak FRS values are generally larger at the top floors than at other floors.

To achieve more accurate predictions of the floor acceleration responses of Chinese buildings, a modified method based on an existing methodology for the generation of FRS was proposed. An empirical fitted formula of the amplification factor was proposed for the resonance region, and it was suggested that the algebraic summation of the first three modes be used to achieve more accurate results. A comparison between the results calculated by the proposed method and THAs demonstrated that the proposed method can provide better FRS predictions for both off-resonance and resonance regions considering different nonstructural damping ratios.

In addition, this study focused on the problem of insufficient consideration of long-period components in existing FRS for seismic performance tests of NSCs. Inelastic time history analysis of standard frame structures under different earthquake intensities was conducted to determine the DAFs of NSCs. FRS for seismic performance tests of NSCs were established based on the analysis results. The proposed response spectra are composed of short period section, linear increase section, platform section, and decrease section.

As a tentative study, several extensions remain to be explored in future studies. First, this study focused solely on RC frames, and the proposed method must be extended to different types of structures (e.g., shear wall structures, passively controlled structures, and base-isolated structures). Second, the nonlinear behaviors of supporting structures under the condition of huge earthquakes should be included in the generation of FRS. In addition, methods for the generation of floor velocity and displacement response spectra are also important when considering the seismic design of different types of NSCs.

REFERENCES

American Society of Civil Engineers. 2016. *ASCE 7-16: Minimum Design Loads for Buildings and Other Structures*. Reston, VI.

Anajafi H. 2018. Improved seismic design of non-structural components (NSCs) and development of innovative control approaches to enhance the seismic performance of buildings and NSCs, Ph.D. Dissertation. University of New Hampshire, Durham.

Applied Technology Council. 1978. *Tentative Provisions for the Development of Seismic Regulations for Buildings*, ATC Report. No. 3-06, Applied Technology Council, Palo Alto, CA.

Aragaw L F, Calvi P M. 2021. Earthquake-induced floor accelerations in base-rocking wall buildings, *Journal of Earthquake Engineering*, **25**(5), 941–969.

CEN. 2004. *Eurocode 8: Design of Structures for Earthquake Resistance-Part 1: General Rules, Seismic Actions and Rules for Buildings*, EN 1998-1. Brussels, Belgium.

D'Angela D, Magliulo G, Cosenza E. 2021. Seismic damage assessment of unanchored nonstructural components taking into account the building response, *Structural Safety*, **93**, 102126.

FEMA P-750. 2009. *NEHRP Recommended Seismic Provisions for New Buildings and Other Structures*. Federal Emergency Management Agency, Washington, DC.

GB 50011-2010. 2010. *Code for Seismic Design of Buildings*. China Architecture and Building Press, Beijing (in Chinese).

IEEE Power Engineering Society. 2005. *IEEE Std 344-2004. Recommended Practice for Seismic Qualification of class 1E Equipment for Nuclear Power Generating Stations.* IEEE Power Engineering Society, New York.

Jiang W, Li B, Xie W C et al. 2015. Generate floor response spectra: Part 1. Direct spectra-to-spectra method, *Nuclear Engineering and Design*, **293**, 525–546.

Masayoshi N, Koji O, Kazuo I. 2010. Generic frame model for simulation of earthquake responses of steel moment frames. *Earthquake Engineering & Structural Dynamics*, **31**(3), 671–692.

Medina R A, Sankaranarayanan R, Kingston K M. 2006. Floor response spectra for light components mounted on regular moment-resisting frame structures, *Engineering Structures*, **28**(14), 1927–1940.

Miyamoto H K, Gilani A S. 2008. Recent earthquakes in Indonesia and Japan: Observed damage and retrofit solutions. In: *Proceeding of 14th World Conference on Earthquake Engineering*, October 12–17, Beijing, China.

NZS 1170.5. 2004. *Structural Design Actions, Part 5: Earthquake Actions*, Draft DR 00902. NZS, New Zealand.

Oropeza M, Favez P, Lestuzzi P. 2010. Seismic response of nonstructural components in case of nonlinear structures based on floor response spectra method, *Bulletin of Earthquake Engineering*, **8**(2), 387–400.

Paulay T, Priestley M J N. 1992. *Seismic Design of Reinforced Concrete and Masonry Buildings*. John Wiley & Sons, Inc, New York.

Pozzi M, Der Kiureghian A. 2015. Response spectrum analysis for floor acceleration, *Earthquake Engineering & Structural Dynamics*, **44**(12), 2111–2127.

Qu Z, Gong T, Li Q et al. 2019. Evaluation of the fishbone model in simulating the seismic response of multistory reinforced concrete moment-resisting frames, *Earthquake Engineering and Engineering Vibration*, **18**(2), 315–330.

Sullivan T J, Calvi P M, Welch D P. 2013. Estimating roof-level acceleration spectra for single storey buildings. In: *4th ECCOMAS Thematic Conference on Computational Methods in Structural Dynamics and Earthquake Engineering*, Kos Island, Greece.

Tian Y, Filiatrault A, Mosqueda G. 2014. Experimental seismic fragility of pressurized fire suppression sprinkler piping joints, *Earthquake Spectra*, **30**(4), 1733–1748.

Vukobratović V, Fajfar P. 2015. A method for the direct determination of approximate floor response spectra for SDOF inelastic structures, *Bulletin of Earthquake Engineering*, **13**(5), 1405–1424.

Vukobratović V, Fajfar P. 2016. A method for the direct estimation of floor acceleration spectra for elastic and inelastic MDOF structures, *Earthquake Engineering & Structural Dynamics*, **45**(15), 2495–2511.

Vukobratović V, Fajfar P. 2017. Code-oriented floor acceleration spectra for building structures. *Bulletin of Earthquake Engineering*, **15**(7), 3013–3026.

Wang T, Shang QX, Li J C. 2021. Seismic force demands on acceleration-sensitive nonstructural components: A state-of-the-art review, *Earthquake Engineering and Engineering Vibration*, **20**(1), 39–62.

Welch DP. 2016. Non-structural element considerations for contemporary performance-based earthquake engineering, Doctoral Dissertation. University of Pavia, Pavia, Italy.

Yasui Y, Yoshihara J, Takeda T, Miyamoto A. 1993. Direct generation method for floor response spectra. In: *Transactions of the 12th international Conference on Structural Mechanics in Reactor Technology (SMiRT 12)*, Stuttgart, Germany, 15–20 August 1993, Paper no K13/4, pp. 367–372.

YD5083-2005. 2006. *Specification for Seismic Test of Telecommunications Equipment.* Beijing University of Posts and Telecommunications Press, Beijing (in Chinese).

4 Shaking Table Tests and Seismic Fragility Analysis of Typical Hospital Equipment

4.1 RESEARCH BACKGROUND

Hospital systems are key strategic buildings in the post-earthquake rescue process, and thus, ensuring post-earthquake hospital functionality to aid individuals injured in earthquakes is critical. However, the severe damage and functionality loss of hospitals have been reported in recent earthquakes [e.g., the earthquakes in Christchurch (New Zealand), 2011; Emilia (Italy), 2012; Lushan (China), 2013; Illapel (Chile), 2015]. In addition to structural damage, the seismic damage of hospital nonstructural components (NSCs), particularly medical equipment (Figure 4.1), usually results in a loss of functionality following an earthquake. NSCs (including inside contents) have been reported to form a significant portion of the total cost of hospital buildings (up to 92%) (Taghavi & Miranda 2003). Thus, the seismic performance of hospital NSCs must be evaluated in order to reduce post-earthquake economic and functionality losses and achieve earthquake-resilient hospital systems.

Several studies on the seismic performance of medical equipment have been conducted over the past decade based on performance-based earthquake engineering. Konstantinidis and Makris (2009) conducted shaking table tests on freestanding laboratory equipment using ground and building floor accelerations recorded during earthquakes. Results revealed the sliding of equipment (up to 600 mm) as the primary response mode of the tested incubator and refrigerator. The authors then conducted probabilistic seismic demand analysis (PSDA) and developed fragility curves of the tested equipment.

Achour (2007) performed shaking table tests on shelves containing medicine and other equipment under several shelf connections (e.g., shelf connected to the structure at the bottom, top and bottom, and bottom and side components). The results indicated the stability of the shelf to increase with the attachment degree. The author also tested nurse tables supported on wheels equipped with locked and unlocked brakes. The unlocked setting was observed to be stable, and the locked setting was observed to be unstable. Kuo et al. (2011) proposed seismic evaluation criteria for the clutter response of medicine shelves via shaking table tests using sinusoidal waves and recorded earthquake motions under three types of medicine shelves.

Cosenza et al. (2015) and Di Sarno et al. (2019) conducted two series of shaking table tests on freestanding hospital cabinets to evaluate their seismic performance

DOI: 10.1201/9781003457459-4

FIGURE 4.1   Damage to hospital equipment in recent earthquakes.

across different parameters. The ground motion was generated based on the AC 156 protocol (ICC-ES 2012), and incremental input earthquake intensities were applied to study the seismic response of cabinets and inside contents. Rocking and overturning limit states were defined for cabinets in Cosenza et al. (2015), and the corresponding fragility curves were developed. In the most recent tests, Di Sarno et al. (2019) defined the overturning and breaking limit states of cabinet contents. Different earthquake intensity measures (IMs) including peak floor acceleration (PFA), peak floor velocity (PFV), and the dimensionless IMs of the PFA and PFV were employed for fragility analysis (Di Sarno et al. 2019). Based on the test results, finite element models were then developed in SAP 2000 to simulate the dynamic properties prior to the rocking of the cabinets (Di Sarno et al. 2015), while rigid block models were adopted to simulate the rocking behavior (Petrone et al. 2016).

Previous research performed full-scale shaking table tests on a four-story RC hospital structure at the E-Defense shaking table facility in Japan to study the seismic performance of both structural and nonstructural components (Sato et al. 2011; Furukawa et al. 2013). The results indicated base isolation as an effective method in reducing the floor acceleration response and preserving the operability and functionality of healthcare services under near-fault ground motions. However, significant sliding and the rocking response of medical equipment supported by casters were observed under long-period ground motions. The vertical peak floor acceleration (PFA_v) was subsequently adopted as an earthquake intensity measure (IM) to evaluate the seismic performance of medical equipment. PFA_v values lower than 1.0 g were associated with few occasions of notable damage, while for PFA_v ranging from 1.0 g to 2.0 g, the jumping and toppling of items on the table, bed, and shelf were observed to occur, and for PFA_v larger than 2.0 g, all items exhibited jumping.

Pantoli et al. (2016) conducted shaking table tests on a full-scale five-story reinforced concrete building furnished with a broad variety of NCSs. Several different types of medical equipment were installed at the fourth- and fifth-floor levels to simulate the layouts of a surgery suite and intensive care unit. The restrained or anchored patient care beds, carts, shelves, and ultrasound imagers showed satisfactory performance to avoid sliding, rolling, or overturning. However, dangerous sliding and toppling of several unrestrained pieces of equipment and pounding with partition walls were observed at a PFA value of 0.56 g.

Although previous studies have provided useful insights into the seismic behavior of freestanding hospital equipment, the majority focus on a single type of floor finishing material, for example, vinyl tile (Konstantinidis and Makris 2009) and linoleum sheet (Cosenza et al. 2015) flooring. However, these floor finishing materials are different from those in Chinese hospitals. In addition, ground motion inputs are typically generated based on the AC 156 protocol (ICC-ES 2012), and the effects of different inputs still require further research. In order to investigate the effects of different earthquake inputs and floor finishing materials on the seismic performance of freestanding hospital cabinets, we conducted shaking table tests of three different types of hospital cabinets. In particular, we adopted three earthquake inputs and two floor finishing materials widely used in Chinese hospitals to evaluate the acceleration, displacement, and rotation responses of cabinets. The test data were used to develop seismic fragility curves of the hospital cabinets corresponding to the different damage states of the cabinet and inside contents.

4.2 EXPERIMENTAL SETUP AND SPECIMENS OF TYPICAL HOSPITAL CABINETS

4.2.1 SHAKING TABLE TEST SETUP AND INSTRUMENTATION

Double-window cabinets were employed for the shaking table tests (Figure 4.2a–c) and divided into three groups: cabinets without drawers (Cabinet A), cabinets with two drawers (Cabinet B), and cabinets with three drawers (Cabinet C). The cabinets were made of cold-formed sheets with dimensions of $1,800\times900\times400$ mm, $1,800\times850\times280/500$ mm, and $1,800\times1,200\times250/500$ mm for Cabinets A, B, and C, respectively. A one-story one-span steel frame (4.35 m $\times$ 4.35 m plan view, 4.2 m column distance, and 2.63 m story height; Figure 4.3) with a 1.2 m $\times$ 2.1 m door on one side of the external wall was used as the loading frame in order to provide floor finishing for the cabinets, simulate the pounding between cabinets and walls, and prevent the cabinets from rushing out of the shaking table under huge earthquake inputs. More details on the loading frame can be found in Dong et al. (2019). The hospital

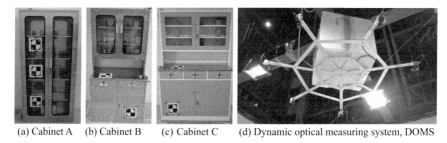

(a) Cabinet A   (b) Cabinet B   (c) Cabinet C   (d) Dynamic optical measuring system, DOMS

FIGURE 4.2 Test hospital cabinets and instrumentation.

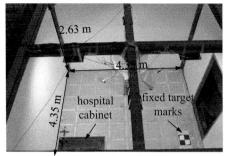
(a) Top-view of the loading frame

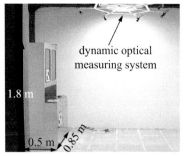
(b) Cabinet B on CMF

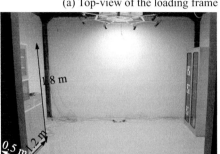
(c) Cabinet A and C on CTF

(d) Cabinet A and C on CMF

FIGURE 4.3 Loading frame and illustration of test setup.

TABLE 4.1
Test Program Definition

Specimen ID	Cabinet Type	Mass (kg)	Floor Finishing Material	Cabinet-to-Wall Distance (mm)
A-CMF-100	Cabinet A	38.8	Cement mortar flooring (CMF)	100
A-CTF-100	Cabinet A	38.8	Ceramic tile flooring (CTF)	100
B-CMF-480	Cabinet B	41.2	Cement mortar flooring (CMF)	480
B-CMF-100-1	Cabinet B	41.2	Cement mortar flooring (CMF)	100
B-CMF-100-2	Cabinet B	41.2	Cement mortar flooring (CMF)	100
C-CMF-100	Cabinet C	51.3	Cement mortar flooring (CMF)	100
C-CTF-100	Cabinet C	51.3	Ceramic tile flooring (CTF)	100

cabinets were tested under cement mortar flooring (CMF) and ceramic tile flooring (CTF). CMF is currently used in some poverty-stricken areas, while CTF is widely used in modern hospitals in China. Two, three, and two types from Cabinets A, B, and C were tested, respectively (Table 4.1). Specimen B-CMF-480 was used to investigate the effects of the cabinet-to-wall distance considering a distance of 480 mm from the adjacent wall. For other specimens, the cabinet-to-wall distance was selected as 100 mm, a representative value in China. The seismic performance assessment of the inside contents is also important for hospital cabinets considering that the seismic damage of contents results in a loss of medical supplies during earthquakes, which can compromise or induce a complete loss of hospital functionality. Thus, material of varying slenderness (e.g., glass bottles, flasks, test tubes, and medicine boxes) was placed inside the cabinets to simulate the actual conditions of a typical hospital room.

High-quality, digital, three-dimensional accelerometers were placed on the shaking table, floor, and cabinets during the shaking table tests to measure the input ground motions, floor acceleration, and responses of the tested cabinets. An accelerometer was attached to the shaking table to record the input acceleration. Furthermore, two accelerometers were used to measure the acceleration response of the top and bottom of the cabinets in group A, while for groups B and C, two accelerometers were used to measure the acceleration response of the top of the cabinet and the platform above the drawers. The displacement and rotation response of the cabinets were also measured. A dynamic optical measuring system (DOMS) was employed to avoid damage to the guyed displacement meters during the overturning of the cabinets. The DOMS system consisted of seven assembled high-resolution industrial cameras (Figure 4.2d). The camera module was installed on the top of the steel frame, and the displacement response of the specimens was determined using the relative coordinate changes between the marks placed on the specimens and those fixed on the floor (Figure 4.3).

4.2.2 Loading Protocol and Test Program

The ground motion-induced damage to the supporting structures will result in changes in the seismic input of the NSCs, e.g., the floor acceleration response, peak floor acceleration (PFA), and floor response spectrum (FRS). Typically, the peak values of FRS/PFA will decrease when the supporting structures suffer larger ground motions and experience heavier seismic damage. The periods corresponding to peak FRS/PFA values will increase as a result of nonlinear behavior of the supporting structures. To consider these effects, seismic performance tests of NSCs are usually conducted using general FRS. In our tests, the general FRSs provided by AC 156 (ICC-ES 2012), Anajafi (2018), and Shang (2021) (Section 3.6) were used to consider these effects. A platform with a long period range is included in all three FRSs, which reflects the possible resonance effect between the NSCs and the supporting structures, as well as the variations of the vibration periods in different buildings. The total length of the input ground motion was set as 40 s. Figure 4.4a–d presents the acceleration time histories and response spectra of three ground motions. Cosenza et al. (2015) and Di Sarno et al. (2019) revealed that the seismic damage of cabinet-type equipment typically occurs in the transversal direction, and thus, we select this as the loading direction for the shaking table tests.

Incremental tests were performed under a peak table acceleration (PTA) of varying intensities (0.05 g, 0.07 g, 0.10 g, 0.14 g, 0.20 g, 0.30 g, 0.40 g, 0.50 g, 0.60 g, 0.70 g, 0.80 g, and 0.90 g). The pounding between the cabinets and floor has an effect on the recorded floor acceleration time histories, resulting in large PFA values that are unable to represent the intensities of the floor seismic input. In order to avoid the distortion of the PFA values, several rounds of seismic shaking were conducted without hospital cabinets on the floor. The recorded PTA and PFA values were employed to perform the regression fitting analysis between the PTA and PFA (Figure 4.4e). The fit between the PTA and PFA was then adopted to determine the IM (i.e., PFA) values in the subsequent analysis.

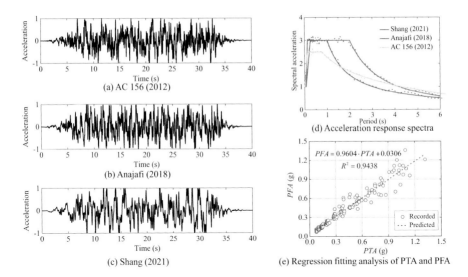

FIGURE 4.4 Input earthquake acceleration time histories and response spectra. (The PGA values are scaled to 1.0 g in a–d.)

4.2.3 HOSPITAL CABINETS

4.2.3.1 Acceleration Response and Acceleration Amplification Factors

The acceleration amplification factor of hospital cabinets is defined as the ratio between the peak component acceleration (PCA) response recorded at the top of the cabinets and the PFA in the loading direction. Figure 4.5 depicts the acceleration amplification factor values for the tested cabinets, where the vertical lines correspond to the PFA values when the cabinets start rocking under different earthquake inputs. The displacement and acceleration amplitude of the response time histories are observed to increase significantly when the cabinets begin to rock. The rocking PFA values can then be determined based on the initial rocking. Prior to the rocking of the cabinet, the amplification factor ranges between 3 and 10. When the rocking begins, the acceleration response increases significantly and at times exceeds the accelerometer measuring range (10.0 g). Furthermore, during rocking, the amplification factor increases with the earthquake intensity, reaching values of 30.0 or higher. Significant changes are observed in the amplification factor pre- and post-rocking. This is attributed to the pounding behavior between the cabinet and floor or adjacent wall during the cabinet rocking process. The pounding behavior results in several spikes recorded in the cabinet acceleration time histories (Figure 4.6), significantly increasing the PCA response of the cabinet.

4.2.3.2 Rotational Angles and Displacement Response

The maximum rotation angles (θ_{max}) of the tested cabinets were determined based on the data from the DOMS system. Prior to rocking, the rotation angles were almost zero and subsequently increased significantly when the tested cabinets started to rock (Figure 4.7). The overturning rotation angles, calculated using the measured maximum values prior to overturning, were determined as 0.194, 0.246, and 0.351 rad

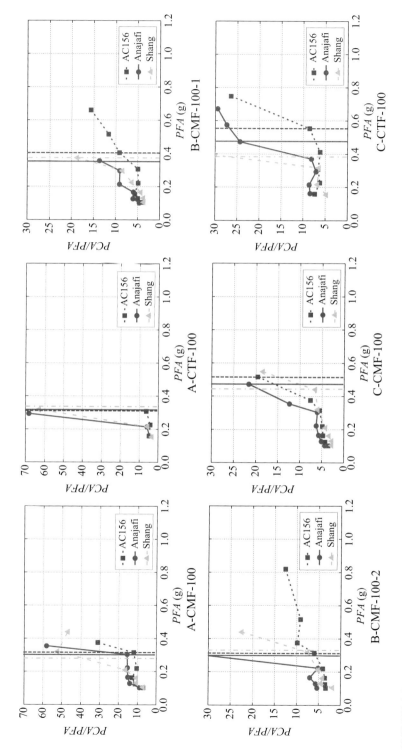

FIGURE 4.5 Acceleration amplification factors for the tested cabinets.

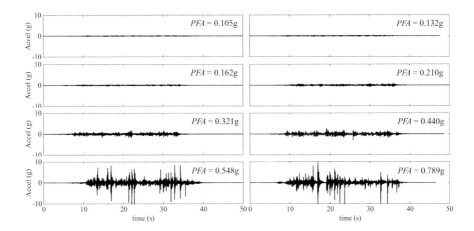

FIGURE 4.6 Acceleration response of cabinet C-CMF-100 under seismic inputs, generated by Shang (2021).

for Cabinets A, B, and C. The configuration of Cabinet A is similar to that of the double-window cabinets tested by Di Sarno et al. (2019), who reported an overturning rotation angle of 0.228 rad. In this study, overturning was observed to occur to Cabinet A (A-CMF-100 and A-CTF-100) for rotation angles equal to 0.228 rad. Furthermore, the maximum rotation angle response varied with the excitation ground motions. The results indicate the maximum rotation angle response under the excitation of the ground motions from Anajafi (2018) and Shang (2021) to generally exceed that of AC 156 (ICC-ES 2012) for similar PFA values. In addition, the effects of floor finishing materials on dynamic rotation response can be demonstrated in Figure 4.7b and c. The effects are obvious for Cabinet C from the perspective of the maximum rotation angle responses. Typically, the rotation angles are usually larger for Cabinet C on CMF than those on CTF, especially for the loading cases near overturning (Figure 4.7c). However, the effects on Cabinet A are not so obvious (Figure 4.7b).

Figure 4.8 shows the final positions and initial cabinet-to-wall distances of the hospital cabinet edges. The photographs of cabinets here represent the final positions of cabinets after loading. Under low earthquake intensities, the cabinets exhibited slight rocking, corresponding to a rotation about the base corners of the cabinets without sliding behavior. The cabinet displacement responses oscillated around the zero position. However, apart from the rocking response, the sliding response was observed for increasing earthquake intensities. The cabinets slid away from the initial position and moved adjacent to the external wall. This phenomenon can be demonstrated clearly by comparing the displacement response of the hospital cabinets under different earthquake intensities (Figure 4.9). Taking Cabinet B-CMF-100-2 as an example, rocking began under a PFA equal to 0.303 g, with a maximum rocking amplitude of 51.7 mm following the seismic inputs generated by Anajafi (2018). The sliding response for this case was almost zero. At the PFA value of 0.354 g, the maximum rocking amplitude increased to 71.6 mm, and the cabinet began to slide with a residual displacement of 34.0 mm. The maximum rocking amplitude was observed to

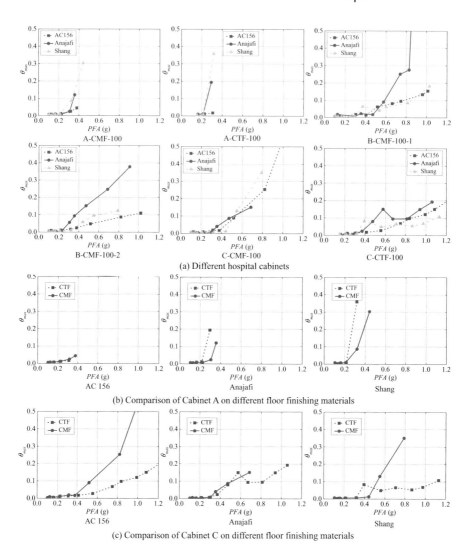

FIGURE 4.7 Maximum rotation angle of hospital cabinets.

increase to 112.4 mm opposite to the external wall under the PFA of 0.472 g, and the residual displacement was enhanced to 74 mm. At the PFA equal to 0.686 g, the maximum rocking amplitude increased to 177.7 mm, and pounding between the cabinet and external wall occurred. Upon the termination of the tests, the gap between the cabinet and the external wall was almost non-existent (Figure 4.9). At the PFA value of 0.907 g, the Cabinet B-CMF-100-2 exhibited an early overturn. Figure 4.10 presents the residual sliding displacements of the tested hospital cabinets. The residual sliding displacements increase with the earthquake intensity, and for the same PFA values, the sliding responses are larger under the excitations generated by Anajafi (2018) and Shang (2021) compared to those of AC 156 (ICC-ES 2012). The response of the hospital cabinets observed in this study differs from the test results reported in

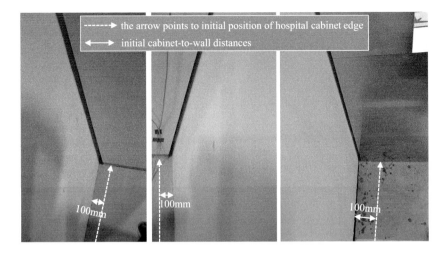

FIGURE 4.8 Final position of hospital cabinet edge.

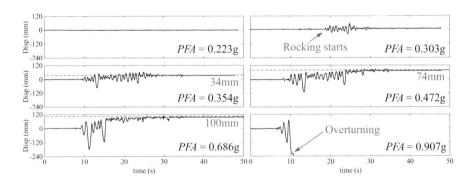

FIGURE 4.9 Displacement response of Cabinet B-CMF-100-2 under the excitation generated by Anajafi (2018).

Cosenza et al. (2015) and Di Sarno et al. (2019), who indicated that sliding is slight or negligible for double- and single-window cabinets that experience rocking.

Note that the partial dissipation of the seismic input energy can occur following the pounding between the cabinets and floor slab (or external wall) during the rocking process. This can effectively prevent the premature overturning of the hospital cabinets. However, the inside contents may suffer from sliding, overturning, and breaking responses when such pounding occurs, particularly for the glass contents (e.g., glass bottles, flasks, and test tubes).

4.2.3.3 Seismic Fragility Evaluation of Hospital Cabinets

The damage experienced by the hospital cabinets and inside contents during the shaking table tests was recorded for each PFA value. Four damage states (DS) were subsequently defined for the hospital cabinets and inside contents: DS1 denotes the sliding of the inside contents; DS2 denotes the rocking of the hospital cabinets;

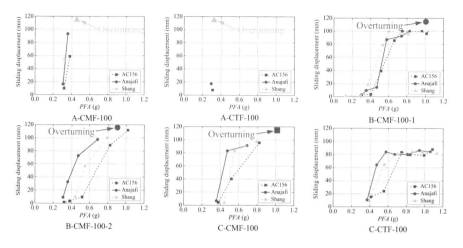

FIGURE 4.10   Residual sliding displacement of hospital cabinets.

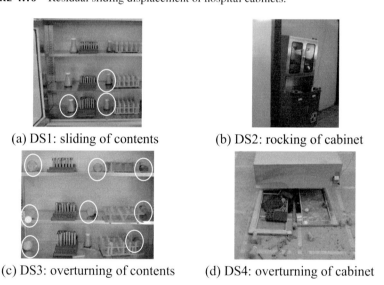

(a) DS1: sliding of contents   (b) DS2: rocking of cabinet

(c) DS3: overturning of contents   (d) DS4: overturning of cabinet

FIGURE 4.11   Exemplary photographs of different damage states.

DS3 denotes the overturning or breaking of the inside contents; and DS4 denotes the overturning of the hospital cabinets (Figure 4.11). The occurrence of DS1 and DS3 is determined by comparing the final position of the inside contents with the initial position during post-shaking inspection and is related to sliding, overturning, or breaking of at least one item of the contents. The occurrence of DS2 is determined by observing the specimen response during testing followed by verification via video recordings and the acceleration and displacement response histories. The occurrence of DS4 is obvious and can be easily determined during the shaking table tests. Note that if DS3 fails to occur prior to the overturning of the cabinet, DS3 and DS4 will occur simultaneously when overturning of the hospital cabinet occurs. Therefore, the overturning of the cabinets represents an upper bound for the sliding, overturning, or breaking of the inside contents.

The seismic fragility curves describe the probability that a component reaches or exceeds damage state dm given a particular engineering demand parameter (EDP, PFA in this study) value and is typically determined using the lognormal distribution given in Equation (4.1) (Porter, Kennedy and Bachman 2007):

$$F_{dm}(\text{edp}) \equiv P[DM \geq dm|EDP = edp] = \Phi\left(\frac{\ln(\text{edp/PFA}_m)}{\beta} \right), \qquad (4.1)$$

where PFA_m is the median value of the peak floor acceleration and β is the logarithmic standard deviation.

We adopt Method A described by Porter, Kennedy and Bachman (2007) to generate the fragility curves for DS1–DS3, while Method C is selected to generate those of DS4 as the overturning data are not sufficient for Method A. In particular, Method A applies to the cases where the damage state is observed for all tested specimens, while Method C applies to those cases where no tested specimens reached the damage state. It should be noted that only the test results of cabinets with a cabinet-to-wall distance of 100 mm were used in seismic fragility analysis. In addition, the cabinets with different cabinet-to-wall distances can be divided into different groups. Then, the fragility curves can be developed for each group using the methods presented in this study.

4.2.3.3.1 Sliding Fragility of the Inside Contents

The sliding of the inside contents is generally the first instance of damage observed in the tests (Figure 4.11a). For the three ground motions used in the shaking table tests, the minimum PFA values corresponding to DS1 of each specimen are used to calculate the values of PFAm and β. Figure 4.12 depicts the seismic fragility curves and

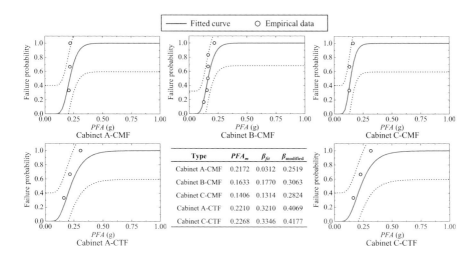

FIGURE 4.12 Sliding fragility curves of the hospital cabinet contents (DS1).

corresponding fragility parameters. Empirical data in Figure 4.12 represent the probabilities of reaching or exceeding the considered damage state and corresponding PFA values obtained by shaking table tests. Sample cumulative distribution (stepped curve) is adopted to generate the empirical fragility data, which is widely used in fragility analysis (Porter, Kennedy and Bachman 2007). We compare the curves of Cabinet A-CMF with Cabinet A-CTF, and Cabinet C-CMF with Cabinet C-CTF to demonstrate the influence of the floor finishing materials on the sliding response of the inside contents. The median PFA values of inside contents begin to slide for Cabinet A-CMF and Cabinet C-CMF, which are 0.2172 g and 0.1406 g, respectively. The corresponding values for Cabinet A-CTF and Cabinet C-CTF are 1.75% and 61.31% larger, respectively. The effects of the ceramic tile flooring compared to the cement mortar flooring on the sliding response of the inside contents of Cabinet C are obvious for Cabinet C. The results also reveal the sliding of the inside contents under low PFA values, which may result in a loss of medical supplies (particularly glass products) under earthquakes.

4.2.3.3.2 Rocking Fragility of the Hospital Cabinets

Figure 4.13a presents the rocking fragility of the hospital cabinets and corresponding fragility parameters. The differences in median PFA values between Cabinet A-CMF and Cabinet A-CTF, and Cabinet C-CMF and Cabinet C-CTF are determined as

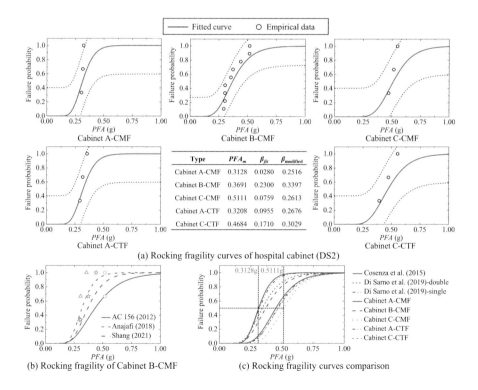

(a) Rocking fragility curves of hospital cabinet (DS2)

Type	PFA_m	β_{fit}	$\beta_{modified}$
Cabinet A-CMF	0.3128	0.0280	0.2516
Cabinet B-CMF	0.3691	0.2300	0.3397
Cabinet C-CMF	0.5111	0.0759	0.2613
Cabinet A-CTF	0.3208	0.0955	0.2676
Cabinet C-CTF	0.4684	0.1710	0.3029

(b) Rocking fragility of Cabinet B-CMF

(c) Rocking fragility curves comparison

FIGURE 4.13 Rocking fragility curves of hospital cabinets (DS2).

2.56% and 8.35%, respectively. The rocking PFA values of the cabinet are similar across the floor finishing materials. However, the type and geometrical configuration of the hospital cabinet exert a great influence on the rocking PFA values. The PFA values corresponding to the rocking of Cabinet C are obviously larger than those of Cabinet A and Cabinet B. Under the same PFA values, the rocking probability of Cabinet A and Cabinet B exceeds that of Cabinet C.

The rocking PFA values under different ground motions are also compared. Results indicate that in general, the PFA values that induce cabinet rocking under the AC 156 (ICC-ES 2012)-generated ground motion surpass those generated by Anajafi (2018) and Shang (2021). For example, the rocking PFA values of Cabinet C-CTF are determined as 0.5530 g, 0.4730 g, and 0.3930 g for AC 156 (ICC-ES 2012), Anajafi (2018), and Shang (2021). Furthermore, the responses under Anajafi (2018) and Shang (2021) are relatively on the conservative side compared to those of AC 156 (ICC-ES 2012). Three specimens in the Cabinet B group were tested under cement mortar flooring (Cabinet B-CMF-480, B-CMF-100-1, and B-CMF-100-2; Table 4.1), and the results are used to provide insights into the effects of the different ground motions. Figure 4.13b presents the rocking fragility curves of Cabinet B located on cement mortar flooring under different earthquake inputs. The median rocking PFA values are determined as 0.4320 g, 0.3162 g, and 0.3682 g for AC 156 (ICC-ES 2012), Anajafi (2018), and Shang (2021), respectively. These results demonstrate the importance of considering different earthquake inputs and frequency contents when generating seismic fragility curves of freestanding cabinets, as suggested by Di Sarno et al. (2019).

Cosenza et al. (2015) reported a median rocking PFA of 0.450 g, while equivalent values were determined as 0.464 g and 0.342 g for single- and double-window hospital cabinets by Di Sarno et al. (2019), with both studies using seismic inputs generated by AC 156 (ICC-ES 2012). Figure 4.13c compares the rocking fragility curves of Cosenza et al. (2015), Di Sarno et al. (2019), and those of this study. Ishiyama (1982) determined a minimum static acceleration ($a_{minimum}$) required to let the rigid block rock as gb/h, where b is the horizontal distance from the center of gravity to the base edge of a body at rest, h is the height of the center of gravity from the base of the body at rest, and g is the gravitational acceleration. The corresponding $a_{minimum}$ values are 0.259 g and 0.232 g for single- and double-window cabinets in tests by Cosenza et al. (2015) and Di Sarno et al. (2019). The $a_{minimum}$ values for Cabinets A, B, and C of this study are 0.222 g, 0.278 g, and 0.278 g, respectively. The shaking table test results indicate that for all the cabinets, these $a_{minimum}$ values are on the safe side. This is mainly due to the cabinet elastic behavior, sliding, and pounding behavior recorded during the tests. Apart from the geometrical configurations of these cabinets, the differences between median PFA values from Cosenza et al. (2015) and Di Sarno et al. (2019) and the current studies are mainly owing to the different seismic inputs. This is further demonstrated by the results that the median rocking PFA value of Cabinet B-CMF under seismic inputs generated by AC 156 (ICC-ES 2012) is 0.4320 g, which is compatible with that from results adopted by using inputs generated by AC 156 (ICC-ES 2012) in studies of Cosenza et al. (2015) and Di Sarno et al. (2019). Compared with the inputs generated by AC 156 (ICC-ES 2012), the inputs generated according to Anajafi (2018) and Shang (2021) have higher platform spectra values and longer platform periods (Figure 4.4). The differences between the inputs result in higher probability of cabinet damage when inputs generated according to Anajafi (2018) and Shang (2021) were used for shaking table tests. This phenomenon can also be found for DS3 and DS4. Therefore, the seismic fragility curves generated in this study provide a conservative estimation of the median PFA values required to meet a given damage state, which is a positive feature for performance tests of NSCs.

4.2.3.3.3 Overturning Fragility of the Inside Contents

The overturning fragility of the inside contents during an earthquake may result in the serious loss of medical supplies, particularly for glass products. Figure 4.14a presents the seismic fragility curves determined based on the recorded PFA values. Note that DS3 and DS4 occurred simultaneously for Cabinet A-CTF, and therefore, the fragility curves of DS3 and DS4 are equal. The median PFA values for DS3 of Cabinet C are obviously larger than those of Cabinets A and B. Moreover, under the same earthquake intensity, the overturning probability of the inside contents in Cabinet C is much lower than that of Cabinets A and B. Di Sarno et al. (2019) determined median PFA values corresponding to the overturning and breaking of the inside contents of hospital cabinets as 0.65 g and 1.18 g, respectively. For Cabinet C, the median PFA values corresponding to the overturning of the inside contents are compatible with that from Di Sarno et al. (2019). However, that value is significantly larger than the results of Cabinets A and B (Figure 4.14b). This is mainly due to the fact that overturning of contents was usually caused by severe cabinet rocking. In

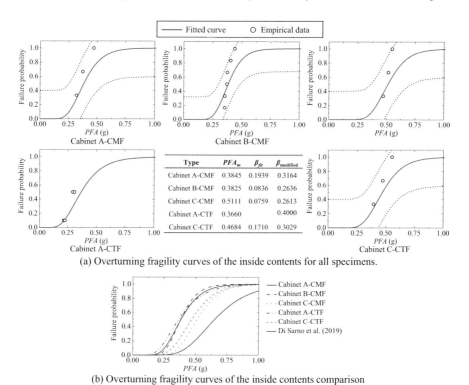

(a) Overturning fragility curves of the inside contents for all specimens.

(b) Overturning fragility curves of the inside contents comparison

FIGURE 4.14   Overturning fragility curves of the hospital cabinet contents (DS3).

addition, Cabinets A and B will start rocking under lower PFA values than the cabinets in Di Sarno et al. (2019) (Figure 4.13c). Moreover, Di Sarno et al. (2019) adopted all content overturning data to generate the DS3 fragility curves without distinguishing single- and double-window cabinets.

4.2.3.3.4 Overturning Fragility of the Hospital Cabinets

As Method A is not able to generate fragility curves based on the overturning data of the hospital cabinets, we employed Method C described by Porter, Kennedy and Bachman (2007) to perform this function. Figure 4.15a presents the fragility analysis results of the overturning damage state for the hospital cabinets. Cabinet A is observed to be the most vulnerable among the tested three hospital cabinet types, regardless of the floor finishing material. All cabinets listed in Table 4.1, with the exception of Cabinet C-CTF-100, overturned during the shaking table tests. Cabinet B-CMF-480 overturned earlier at a PFA value of 0.3043 g, while the other two specimens of Cabinet B (B-CMF-100-1 and B-CMF-100-2) overturned at PFA values of 1.01 g and 0.907 g, respectively. Note that the only difference between these three cabinets is the cabinet-to-wall distance. This agrees with the observations reported in previous research (Filiatrault, Kuan and Tremblay 2004; Cosenza et al. 2015; Di Sarno et al. 2019) that indicate the presence of the wall to be effective in preventing the overturning of cabinets. This is attributed to the partial dissipation of the seismic input energy as a result of the pounding between the cabinets and wall. The reduction

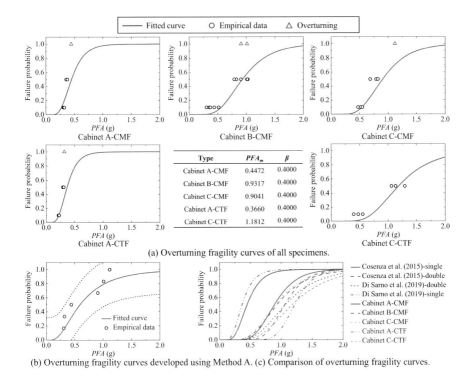

(a) Overturning fragility curves of all specimens.

(b) Overturning fragility curves developed using Method A. (c) Comparison of overturning fragility curves.

FIGURE 4.15 Overturning fragility curves of hospital cabinets (DS4).

in the cabinet-to-wall distance can significantly improve the seismic performance of hospital cabinets when the overturning limit state is considered.

The overturning PFA values are employed to develop seismic fragility curves based on Method A without distinguishing the cabinet types and floor finishing material (Figure 4.15b). The median PFA value and β-fit are determined as 0.5931 g and 0.599 g, respectively, indicating a large dispersion. The results are conservative for Cabinets B and C, but not for Cabinet A. Considering that the cabinet types are not distinguished when generating fragility curve in Figure 4.15b, this curve is not suitable for direct application in seismic performance assessments. Figure 4.15c compares the observations in this study with those reported in Cosenza et al. (2015) and Di Sarno et al. (2019). The overturning fragility curves of Cabinets B and C in this study are similar to those of the double- and single-window cabinets in the previous research. However, Cabinet A overturns at a relatively small earthquake intensity (PFA) compared to the previous results.

Figure 4.16 depicts the seismic fragility curves corresponding to the four damage states of the hospital cabinets. The occurrence probability that a cabinet is in damage state dm, $P[DM = dm|EDP = edp]$, can be calculated using Equation (3.2):

$$P[DM = dm|EDP = edp] = \begin{cases} 1 - F_1(edp) & dm = 0 \\ F_{dm}(edp) - F_{dm+1}(edp) & 1 \le dm < 4, \\ F_{dm}(edp) & dm = 4 \end{cases} \quad (3.2)$$

where dm = 0 represents no damage. Note that $P[DM = dm|EDP = edp]$ can assume negative values at times, for example, $P[DS2|PFA = 0.55\text{ g}]$ equals -0.036 for Cabinet B-CMF, which is meaningless. In order to overcome this problem, we implement Equation (3.3) to modify the seismic fragility curves:

$$F_i(edp) = \max_j \left\{ \Phi(\frac{\ln(edp / PFA_{mj})}{\beta_j}) \right\} \quad \text{for all } j \ge i. \quad (3.3)$$

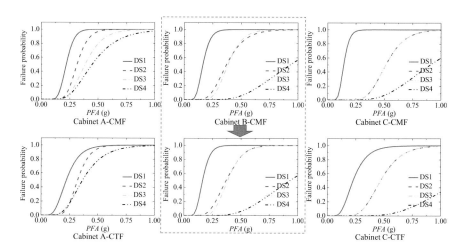

FIGURE 4.16 Seismic fragility curves corresponding to the four damage states.

Figure 4.16 presents the modified seismic fragility curves for Cabinet B-CMF. The developed seismic fragility curves of the hospital cabinets considering different cabinet types and floor finishing materials can form a part of the seismic performance assessment and seismic resilience evaluation of hospital buildings. The seismic loss (e.g., economic loss resulting from damage of hospital cabinets and inside contents) can then be determined based on the total probability theory using the occurrence probability and local norm cost values of cabinets and inside contents, as described in Equation (3.4),

$$\text{Loss} = \sum_{dm=1}^{4} L_{dm} \cdot P[\text{DM} = \text{dm}|\text{EDP} = \text{edp}], \tag{3.4}$$

where L_{dm} is the local norm cost value corresponding to damage state dm.

4.3   INFUSION SUPPORTS

Freestanding infusion supports are still widely used in hospitals in remote areas of China, and most of these hospitals are low-rise buildings. A total of six types of infusion supports were investigated in this chapter, as shown in Figure 4.17a. There were three specimens of the same configuration for each type of infusion support. Four bottles filled with water to simulate the real infusion state were installed on top of each infusion support. The distance between the top of infusion support and the floor slab was 1.85 m. In order to avoid the mutual influence when the infusion support overturns and makes full use of the test space, the arrangement scheme of infusion supports in each loading case is shown in Figure 4.17b.

During the shaking table tests of infusion supports, the artificial ground motions developed based on FRS of AC 156 (ICC-ES 2012), Anajafi (2018), and Shang (2021) were used for tests of Type 1. The input PTAs of 0.05 g, 0.07 g, 0.10 g, 0.14 g, and 0.20 g was considered. The seismic responses of Type 1 infusion supports are listed in Table 4.2.

The maximum residual displacements of three same specimens were selected to represent the results of each loading case, as shown in Figure 4.18a. It can be found that the response of infusion supports under excitation of AC156 (2012) is the smallest, while that under the excitation of Shang (2021) (Section 3.6) is the largest. The relationship between the maximum residual displacement (*ResidualDisp*) and PFA is shown in Figure 4.18b. The corresponding seismic fragility curves of infusion support (Type 1) are shown in Figure 4.18c. The same trend can be found in that the failure probability under the excitation of Shang (2021) is much larger than that of AC156 (2012) and Anajafi (2018). In addition, it should be noted that the logarithm standard deviation of the fragility curve by Anajafi (2018) reached 0.897 due to the fact that the residual displacement response distribution is relatively discrete. If we consider the test results under three different input motions, the developed seismic fragility curve is the gray curve in Figure 4.18c. Based on this curve, the seismic fragility curves by AC156 (2012) and Anajafi (2018) will significantly underestimate the failure probability, while the motion by Shang (2021) can develop a relatively more conservative seismic fragility curve and can satisfy the seismic performance

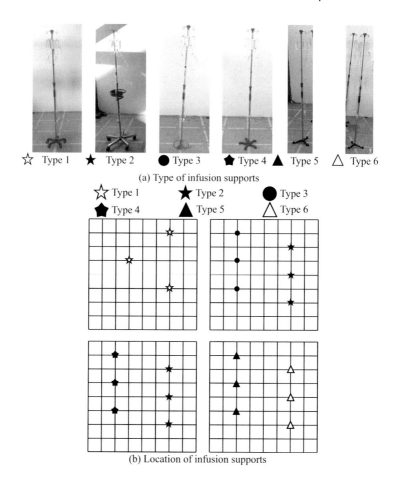

(a) Type of infusion supports

(b) Location of infusion supports

FIGURE 4.17 Layout plan of infusion supports.

test requirement. Therefore, only the artificial motion generated according to the FRS proposed in this study by Shang (2021) is used for loading in subsequent tests.

For shaking table tests of freestanding infusion supports, all Type 1 infusion supports overturned. Only one of the Type 2 infusion supports overturned under excitation of 0.30 g, and the other two specimens of Type 2 didn't overturn after excitation of 0.60 g. All Type 3 infusion supports overturned under excitation of 0.14 g, while the three Type 4 infusion supports overturned under excitation of 0.20, 0.30, and 0.40 g, respectively. Considering that the dimensions and response of Types 5 and 6 are basically the same, the six specimens are considered as one group. Two of them overturned under excitation of 0.07 g, three of them overturned under excitation of 0.10 g, and one of them overturned under excitation of 0.20 g. In summary, the sliding of the wheels of the infusion supports can effectively avoid overturning. However, when the sliding distance reaches a certain limit in the actual use in hospitals, the pounding with other objects (such as hospital beds, tables, and chairs) will still induce the risk of overturning or affect the normal use (such as bottles broken).

TABLE 4.2

Response of Type 1 Infusion Supports

Input Motions	PTA	Residual Displacement/mm			Observed Response
		1	2	3	
AC156	0.05 g	–	–	–	Bottles wobbled slightly
	0.07 g	–	–	–	Bottles wobbled and rocking of infusion support
	0.10 g	30	–	–	Bottles wobbled and rocking of infusion support
	0.14 g	20	20	30	Bottles wobbled and rocking of infusion support
	0.20 g	180	46	40	Bottles wobbled and rocking of infusion support
Anajafi	0.05 g	–	–	–	Bottles wobbled slightly
	0.07 g	–	–	–	Bottles wobbled and rocking of infusion support
	0.10 g	50	–	–	Bottles wobbled and rocking of infusion support
	0.14 g	100	25	22	Bottles wobbled and rocking of infusion support
	0.20 g	255	61	40	Bottles wobbled and rocking of infusion support
Proposed (Shang 2021)	0.05 g	–	–	–	Bottles wobbled heavily
	0.07 g	–	–	–	Bottles wobbled heavily and rocking of infusion support
	0.10 g	60	40	10	Bottles wobbled heavily and rocking of infusion support
	0.14 g	80	30	20	Tend to overturn of infusion support
	0.20 g	201	89	54	Almost overturn of Specimen 3

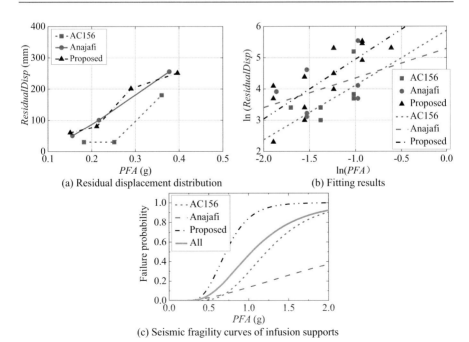

(a) Residual displacement distribution
(b) Fitting results
(c) Seismic fragility curves of infusion supports

FIGURE 4.18 Residual displacement and seismic fragility curves of Type 1 infusion support. *Note*: "all" represents the test results under three different input motions.

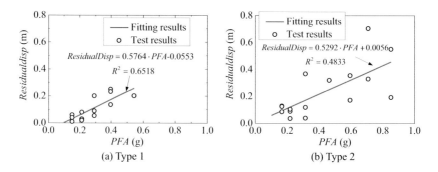

FIGURE 4.19 Residual displacement of Types 1 and 2.

The relationship between the maximum residual displacement and PFA of Types 1 and 2 is shown in Figure 4.19. The seismic damage state of infusion supports can be defined as overturning or the residual displacement reaches 300 mm. Then, the seismic fragility curves of different types of infusion supports can be developed as shown in Figure 4.20, and the corresponding seismic fragility parameters are listed in Table 4.3. The results show that Types 5 and 6 are the most prone to overturning, followed by Types 2 and 4, and Type 1 has the best seismic performance. If the type of infusion supports is not distinguished, the damage and damage data of all infusion stands can be fitted to obtain the seismic fragility curve of all infusion supports, as shown in Figure 4.20e. When the PFA is about 0.28 g, the exceedance probability of infusion supports failure is 50%.

4.4 MEDICAL RESUSCITATION CARTS

Two types of medical resuscitation carts were tested in this study, i.e., simple medical resuscitation carts and multi-purpose medical resuscitation carts. For simple medical resuscitation carts, CMF was used, while both CMF and CTF were considered for multi-purpose medical resuscitation carts.

4.4.1 SIMPLE MEDICAL RESUSCITATION CARTS

There were three simple medical resuscitation carts in the tests, and two configurations of brake relaxed and brake locked were considered. The main response of the simple medical resuscitation carts under earthquake inputs was sliding (brake-locked state) and rolling (brake-relaxed state). The distance of the simple medical resuscitation cart from the initial position to the final position after tests was defined as residual displacement. The residual displacements under different ground motion intensities and different seismic inputs are shown in Figure 4.21b and c. From the residual displacement response of the simple medical resuscitation carts and the phenomena observed and recorded in the tests, the response is the smallest when the excitation of AC156 (2012) was adopted, followed by Anajafi (2018), and the maximum response can be found for the proposed FRS under similar intensities.

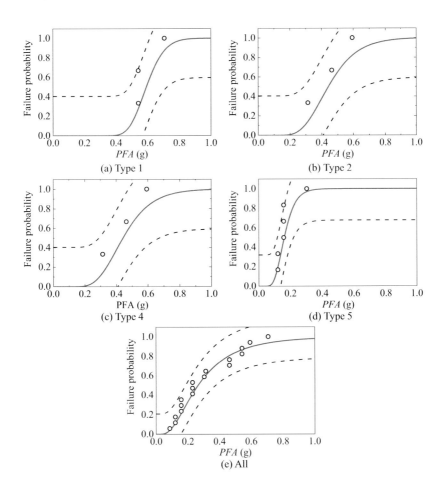

FIGURE 4.20   Seismic fragility curves of infusion support.

TABLE 4.3
Fragility Parameters of Infusion Supports

Type	PFA$_m$ (g)	β
Type 1	0.5926	0.1525
Type 2	0.4417	0.3216
Type 4	0.4417	0.3216
Types 5 and 6	0.1606	0.3388
All	0.2779	0.6390

Based on the existing damage research on equipment supported on wheels/casters (Nikfar & Konstantinidis 2017), overturning will not occur. In this study, the seismic damage state of these equipment is defined as exceeding a given displacement limit. Probabilistic seismic demand analysis and probabilistic seismic

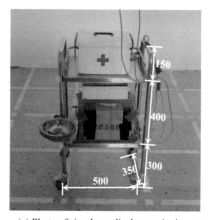

(a) Photo of simple medical resuscitation carts

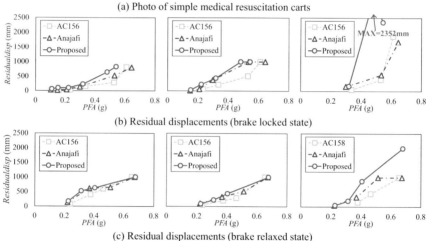

(b) Residual displacements (brake locked state)

(c) Residual displacements (brake relaxed state)

FIGURE 4.21 Simple medical rescue cart.

fragility analysis are used to establish the seismic fragility model. In order to increase the number of samples used for probabilistic seismic demand analysis, the responses under different ground motion inputs will be collected as the same group. Only the two configurations of brake relaxed and brake locked were distinguished. The probabilistic seismic demand models are shown in Figure 4.22a. It can be seen that keeping the brake relaxed or locked has little influence on the residual displacement response of a simple medical rescue cart. The corresponding seismic fragility curves are shown in Figure 4.22b, where D_c is the limit value of residual displacement, which is respectively selected as 0.1, 0.5, 1.0, 1.5, 2.0, and 3.0 m. It can be seen that keeping the brake relaxed or locked has little influence on the seismic fragility curves, and this result is the same as that of Nikfar and Konstantinidis (2017).

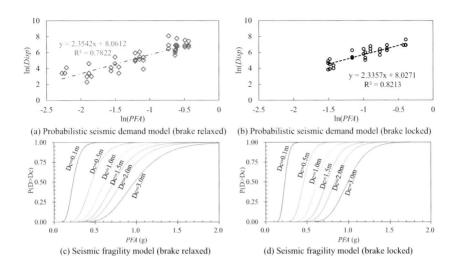

(a) Probabilistic seismic demand model (brake relaxed)　　(b) Probabilistic seismic demand model (brake locked)

(c) Seismic fragility model (brake relaxed)　　　　(d) Seismic fragility model (brake locked)

FIGURE 4.22　Fragility analysis results of simple medical rescue cart.

4.4.2　MULTI-PURPOSE MEDICAL RESUSCITATION CARTS

In the test, the multi-purpose medical resuscitation carts considered two configurations (brake relaxed and brake locked), and different floor finishing materials including CMF and CTF are considered, as shown in Figure 4.23a. The movement tracks of the multi-purpose medical resuscitation carts are shown in Figure 4.23b and c. It could be found that although the shaking table input in the test was unidirectional loading, its response is multidirectional due to the asymmetry of the multi-purpose medical resuscitation cart itself. The sliding (brake-locked) or rolling (brake-relaxed) response is bidirectional instead of unidirectional sliding or rolling in the loading direction. In addition, the sliding and rolling trajectory curves are obviously different. For brake-relaxed cases, the multi-purpose medical resuscitation carts reciprocating oscillated along the direction of the loading while rolling in a small area perpendicular to the loading direction. For brake-locked cases, the multi-purpose medical resuscitation carts basically slid in one direction. The maximum displacements under different ground motion intensities and different seismic inputs are shown in Figure 4.23d and e. According to the maximum displacement response of the multi-purpose medical resuscitation carts and the phenomena observed and recorded in the test, the response is the smallest under the excitation of AC156 (2012), followed by Anajafi (2018). The maximum response is under the excitation of the proposed FRS (Shang 2021).

The probabilistic seismic demand models of multi-purpose medical rescue carts are shown in Figure 4.24a, c, e, and f. It can be found that for this kind of medical rescue carts, the brake relaxed or locked and the floor finishing materials will have a great influence on the seismic response. Specifically, for the CTF floor under the excitation with PFA less than 0.2 g, the sliding distance is small, and the sliding displacements under brake-locked and brake-relaxed conditions are almost the same. However, when the PFA is greater than 0.2 g, the sliding distance under brake-relaxed conditions will increase significantly and is much larger than the sliding displacements under brake-locked conditions. For the CTF floor, when the PFA is less than 0.4 g, the sliding displacement under the brake-relaxed conditions is slightly larger than that under the brake-locked conditions but the difference is not so large. When the PFA is greater than 0.4 g, the sliding displacement under the brake-locked conditions is significantly larger than that under the brake-relaxed conditions.

(a) Photo of multi-purpose medical resuscitation carts

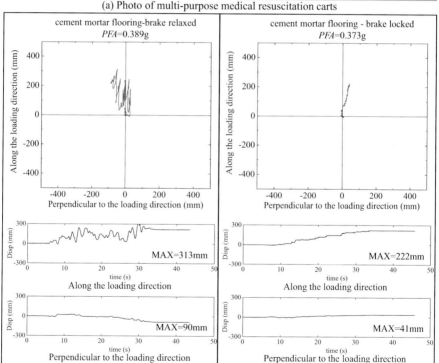

(b) Movement tracks (CMF)

FIGURE 4.23   Multi-purpose medical rescue carts.

(Continued)

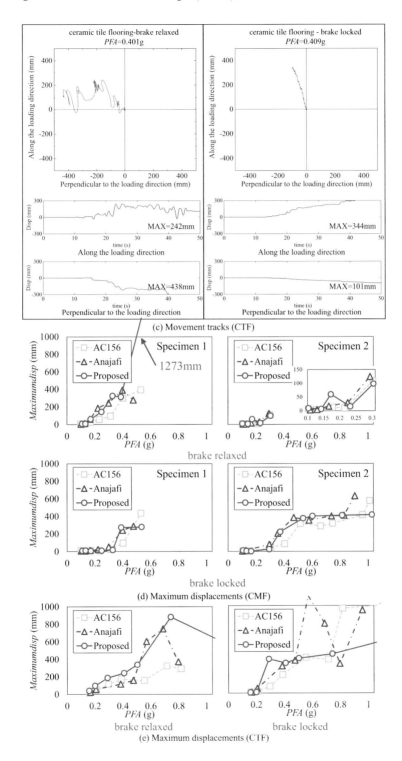

FIGURE 4.23 (*Continued*) Multi-purpose medical rescue carts.

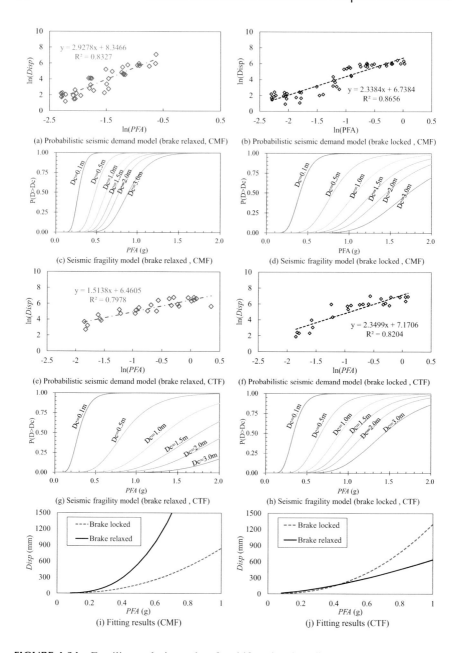

FIGURE 4.24 Fragility analysis results of multifunctional medical rescue cart.

The seismic fragility curves of multi-purpose medical rescue carts are shown in Figure 4.24b and d. For the CMF, locking the brake can effectively reduce the probability that the sliding displacement of medical rescue carts exceeds the displacement limit. For the CTF, when D_c is small (such as 100 mm), the sliding displacement of medical rescue carts is more likely to exceed the displacement limit under the

TABLE 4.4

Sliding Fragility Parameters of Medical Rescue Carts

Floor Finishing Materials	Medical Rescue Cart Types	Dc (m)	PFA$_m$ (g)						β
			0.1	0.5	1.0	1.5	2.0	3.0	
CMF	Simple medical resuscitation carts	Brake relaxed	0.2304	0.4564	0.6127	0.7278	0.8224	0.9770	0.3036
		Brake locked	0.2311	0.4602	0.6193	0.7367	0.8332	0.9912	0.1880
CMF	Multi-purpose medical resuscitation carts	Brake relaxed	0.2786	0.4828	0.6117	0.7026	0.7752	0.8903	0.2322
		Brake locked	0.4016	0.7993	1.0751	1.2787	1.4461	1.7199	0.2931
CTF	Multi-purpose medical resuscitation carts	Brake relaxed	0.2936	0.8501	1.3437	1.7565	2.1241	2.7765	0.3492
		Brake locked	0.3356	0.6658	0.8942	1.0626	1.2010	1.4271	0.3107

condition of brake relaxed. When D_c is large, the sliding displacement of medical rescue carts is more likely to exceed the displacement limit under the condition of brake locked. The sliding fragility parameters of multi-purpose medical rescue carts obtained from the shaking table test are listed in Table 4.4.

4.5 PATIENT BEDS

Considering that most of the patient beds (or operating beds) are in brake-locked state used in hospitals, three configuration conditions are considered in this study during the shaking table test of medical beds, including brake locked on CMF, brake relaxed on CMF, and brake locked on CTF, as shown in Figure 4.25a. The movement tracks of the patient beds are shown in Figure 4.25b. It can be found that although the shaking table input in the test is unidirectional loading, the sliding response is multidirectional instead of unidirectional sliding along the loading direction due to the asymmetry of the bed itself. It can be further found from the probabilistic seismic demand model in Figure 4.26a, c, and e that the maximum response of the bed on CMF floor is slightly larger than that of the bed on CTF under the condition of brake locked. In addition, the maximum displacement of the hospital bed on the CMF under the brake-relaxed condition is much greater than that under the brake-locked condition, which indicates that keeping the brake locked can effectively reduce the displacement response of the hospital bed. The probabilistic seismic fragility curves of the hospital patient beds are shown in Figure 4.26b and d. Under a given ground motion intensity, the occurrence probability of the maximum displacement of the hospital bed exceeding a given limit under the condition of brake locked is much smaller than that under the condition of brake relaxed.

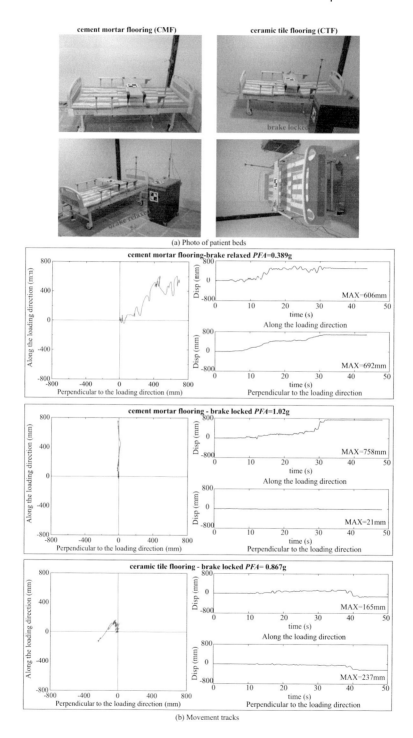

(a) Photo of patient beds

(b) Movement tracks

FIGURE 4.25   Medical patient beds.

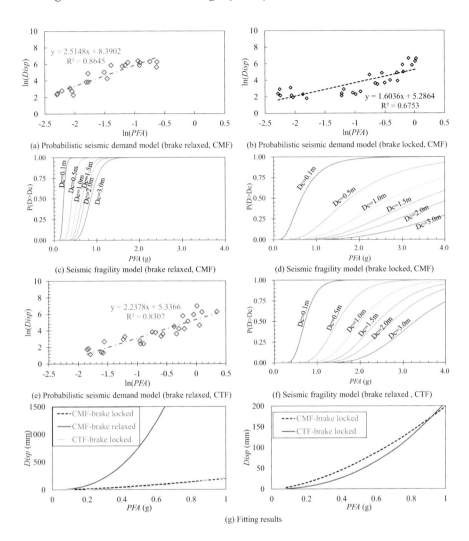

FIGURE 4.26   Fragility analysis results of medical patient beds.

4.6   SHADOWLESS LAMPS

For shaking table tests of medical shadowless lamps, only the CTF was considered, and the shadowless lamp was electrified to check the lighting performance after earthquakes, as shown in Figure 4.27a. For all the loading cases, the shadowless lamps were always in good condition and no damage to the functionality was found. But the shadowless lamps were found to slide and rotate due to asymmetry, and there is no brake locking device. Pounding with the adjacent wall was found during the tests. In a hospital, this type of shadowless lamp may also collide with the nearby medical staff, patients, operating bed, and operation cart. The movement track is shown in Figure 4.27b–d, and the probabilistic seismic demand model and fragility curves are shown in Figure 4.28.

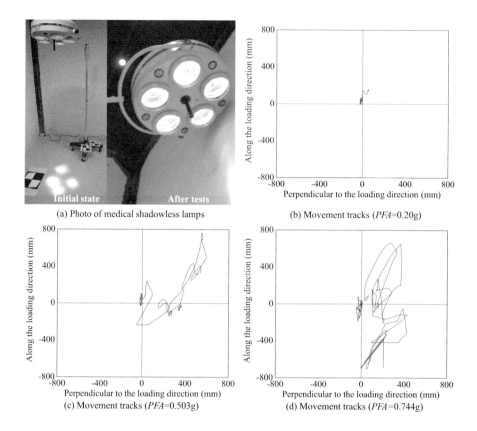

(a) Photo of medical shadowless lamps

(b) Movement tracks (*PFA*=0.20g)

(c) Movement tracks (*PFA*=0.503g)

(d) Movement tracks (*PFA*=0.744g)

FIGURE 4.27   Medical shadowless lamps.

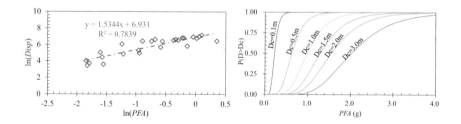

FIGURE 4.28   Fragility analysis results of medical shadowless lamps.

4.7   SUMMARY AND CONCLUSIONS

In this study, we conducted shaking table tests on different hospital equipment widely used in Chinese hospitals. Three artificial ground motions were adopted as the seismic inputs. The seismic responses of the hospital equipment were evaluated, and the seismic fragility curves were developed based on the test results. Furthermore, the effects of different earthquake inputs and floor finishing materials were analyzed. The key results and conclusions can be summarized as follows:

1. The acceleration amplification factor (PCA/PFA) is defined and used to compare the cabinet acceleration response. The amplification factor is relatively small (3.0–10.0) when the cabinet is not observed to rock. However, the acceleration amplitude of the response time histories increases significantly when rocking commences. The acceleration response of the cabinets is also significantly enhanced (up to 30.0). The amplification factor value changes significantly pre- and post-rocking. This is generally attributed to the pounding between the cabinet and floor or adjacent wall during rocking. The pounding behavior results in several spikes in the cabinet acceleration time histories, significantly increasing the PCA response of the cabinet.

2. Although the pounding behavior can dissipate the seismic input energy and effectively avoid the premature overturning of the cabinets, it increases the risk of sliding and overturning of the inside contents. This is not conducive to the seismic protection of the inside contents. The results also indicate the presence of the wall to be effective in preventing the overturning of the hospital cabinets. Reducing the cabinet-to-wall distance can significantly improve the seismic performance of the hospital cabinets.

3. Prior to the occurrence of rocking, the rotation angles are almost zero and subsequently increase significantly when rocking begins. The cabinets exhibit slight rocking under low earthquake intensities, and the cabinet displacement responses oscillate around the zero position. However, apart from the rocking response, the sliding response is observed as the earthquake intensity increases. The observed responses of the hospital cabinets differ from those of the existing studies, indicating that sliding is slight or negligible when rocking occurs for double- and single-window cabinets.

4. The maximum rotation angle response under the excitation of the ground motions from Anajafi and Shang is generally larger than those of AC 156 for similar input PFA values. PFA values that initiate cabinet rocking under the AC 156 generated ground motion exceed those generated by Anajafi and Shang. This highlights the need to consider the effects of different earthquake inputs.

5. The effects of the floor finishing materials on the sliding response of the inside contents in Cabinet C with ceramic tile flooring are obvious compared with cement mortar flooring. However, the rocking PFA values of the cabinet are similar across floor finishing materials.

6. Based on the test results, the sliding and overturning seismic damage limit states were defined for the inside contents, while the rocking and overturning limit states were defined for the hospital cabinets. The type and geometrical configuration of the hospital cabinets are observed to exert a great influence on the cabinet rocking response. In particular, the rocking PFA values of Cabinet C are obviously larger than those of Cabinets A and B, while the median overturning PFA values of contents inside Cabinet C surpass those of Cabinets A and B. Despite the differences in hospital cabinet configurations between this study and existing studies, Cabinets B and C rocking fragility curves exhibit similar median PFA values. The median PFA values corresponding to the overturning and breaking of the inside

contents determined in this study are significantly lower than those reported from existing studies. The seismic fragility curves also revealed Cabinet A to be the most vulnerable among the tested three hospital cabinet types for the four considered damage states.

REFERENCES

Achour N. 2007. Estimation of malfunction of a healthcare facility in case of earthquake, Ph.D. Dissertation. Kanazawa University, Kanazawa, Japan.

Anajafi H. 2018. Improved seismic design of non-structural components (NSCs) and development of innovative control approaches to enhance the seismic performance of buildings and NSCs, Ph.D. Dissertation. University of New Hampshire, NH, USA.

Cosenza E, Di Sarno L, Maddaloni G, Magliulo1 G, Petrone C, Prota A. 2015. Shake table tests for the seismic fragility evaluation of hospital rooms, *Earthquake Engineering & Structural Dynamics*, **44**(1), 23–40.

Di Sarno L, Petrone C, Magliulo G, Manfredi G. 2015. Dynamic properties of typical consultation room medical components, *Engineering Structures*, **100**, 442–454.

Di Sarno L, Magliulo G, D'Angela D, Cosenza E. 2019. Experimental assessment of the seismic performance of hospital cabinets using shake table testing, *Earthquake Engineering & Structural Dynamics*, **48**(1), 103–123.

Dong H Y, Yang L, Cao W L, Lv X M, Bian J L. 2019. Shaking table test of a light steel frame structure with micro-crystalline foam plates, *Journal of Harbin Institute of Technology*, **51**(12), 27–34 (in Chinese).

Filiatrault A, Kuan S, Tremblay R. 2004. Shake table testing of bookcase-partition wall systems, *Canadian Journal of Civil Engineering*, **31**(4), 664–676.

Furukawa S, Sato E, Shi Y, Becker T, Nakashima M. 2013. Full-scale shaking table test of a base-isolated medical facility subjected to vertical motions, *Earthquake Engineering & Structural Dynamics*, **42**(13), 1931–1949.

International Code Council Evaluation Service (ICC-ES). 2012. *AC156: Acceptance Criteria for Seismic Qualification by Shake-Table Testing of Nonstructural Components and Systems*. International Code Council (ICC 2006), International Building Code, Whittier, CA.

Ishiyama Y. 1982. Motions of rigid bodies and criteria for overturning by earthquake excitations, *Earthquake Engineering & Structural Dynamics*, **10**(5), 635–650.

Kaneko M, Hayashi Y. 2004. A proposal for simple equations to express a relation between overturning ratios of rigid bodies and input excitation. In: *Proceedings of the 13th World Conference on Earthquake Engineering (Paper* No. 3299), 1–6 August, Vancouver, BC, Canada.

Konstantinidis D, Makris N. 2009. Experimental and analytical studies on the response of freestanding laboratory equipment to earthquake shaking, *Earthquake Engineering & Structural Dynamics*, **38**(6), 827–848.

Kuo K C, Suzuki Y, Katsuragi S, Yao G C. 2011. Shake table tests on clutter levels of typical medicine shelves and contents subjected to earthquakes, *Earthquake Engineering & Structural Dynamics*, **40**(12),1367–1386.

Nikfar F, Konstantinidis D. 2017. Shake table investigation on the seismic performance of hospital equipment supported on wheels/casters, *Earthquake Engineering & Structural Dynamics*, **46**(2), 243–266.

Pantoli E, Chen M C, Wang X, Astroza R, Ebrahimian H, Hutchinson T C, Conte J P, Restrepo J I, Marin C, Walsh K D, Bachman R E, Hoehler M S, Englekirk R, Faghihi M. 2016. Full-scale structural and nonstructural building system performance during earthquakes: Part II-NCS damage states, *Earthquake Spectra*, **32**(2), 771–794. doi:10.1193/012414eqs017m.

Petrone C, Di Sarno L, Magliulo G, Cosenza E. 2016. Numerical modelling and fragility assessment of typical freestanding building contents, *Bulletin of Earthquake Engineering*, **15**(4), 1609–1633.

Porter K, Kennedy R, Bachman R. 2007. Creating fragility functions for performance-based earthquake engineering, *Earthquake Spectra*, **23**(2), 471–489.

Sato E, Furukawa S, Kakehi A, Nakashima M. 2011. Full-scale shaking table test for examination of safety and functionality of base-isolated medical facilities, *Earthquake Engineering & Structural Dynamics*, **40**(13), 1435–1453.

Shang Q X. 2021. Research on seismic resilience assessment method of hospital systems, Ph.D. Dissertation. Institute of Engineering Mechanics, China Earthquake Administration (in Chinese).

Taghavi S, Miranda E. 2003. Response Assessment of Nonstructural Building Elements, PEER Report 2003/05. Pacific Earthquake Engineering Research Center, Berkeley, CA.

Wang Y M, Xiong L H, Xu W X. 2013. Seismic damage and damage enlightenment of medical buildings in Lushan MS 7.0 earthquake, *Earthquake Engineering and Engineering Dynamics*, **4**(4), 44–53 (in Chinese).

5 A Quantitative Framework to Evaluate the Seismic Resilience of Hospital Systems

5.1 INTRODUCTION

Hospital systems are recognized as strategic buildings in hazardous events and play a key role in disaster rescues, especially after an earthquake. However, many hospitals have reportedly suffered significant damage and lost their functionality during earthquakes (OSHPD 1995; DOHE 1995; CSSC 2000; Myrtle et al. 2005; Moehle et al. 2010; Wang et al. 2013; Malkin and Semple 2017; Santarsiero et al. 2018). For instance, the 1971 San Fernando earthquake caused severe damage to four hospitals (Jennings and Housner 1971). Following the 1994 Northridge earthquake, a study was conducted by the California Office of Statewide Health Planning and Development (OSHPD) to survey the damage to 140 hospitals and 327 nursing homes. Eight nursing homes and 20 hospitals of them were identified as unsafe for occupancy or entry (OSHPD 1995; Myrtle et al. 2005). Similar damage to hospital systems was observed after the 1995 Kobe earthquake (Myrtle et al. 2005; Santarsiero et al. 2018). Nearly 20% of hospitals in the affected area were partially or completely inoperable after the 2010 Chile earthquake (Moehle et al. 2010). During the 2013 M_S 7.0 Lushan earthquake in China, 9 out of 22 county medical buildings experienced moderate to severe damage. The nine buildings were evacuated completely, causing three hospitals to lose their functionality entirely and one hospital to lose 60% of its functionality (Wang et al. 2013). After the 2016 Central Italy earthquake, some small to medium hospitals partially or totally lost their functionality and had to move the inpatients to other hospitals (Santarsiero et al. 2018). More recently, the 2017 Mexico earthquake forced the main hospital to be evacuated after it suffered major damage (Malkin and Semple 2017). In these earthquakes, emergency rescue was strongly affected owing to the reduced functionality of the hospitals, which initiated a comprehensive study to evaluate and improve the seismic performance of the hospital systems.

Among many performance indices of hospital buildings, seismic resilience, which is defined as the ability of an engineering system to resist, restore, and adapt to an earthquake impact (Bruneau et al. 2003; Bruneau and Reinhorn 2007) has caused much concern recently; the recovery capacity is quantified as the variation of the functionality over time, as shown in Figure 5.1. The ability of the system to maintain functionality after earthquakes and that of the system to recover from earthquakes are two essential properties that a system should possess before or after the occurrence

 DOI: 10.1201/9781003457459-5

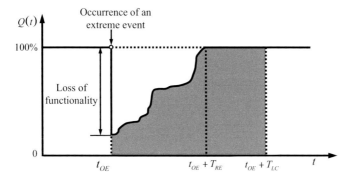

FIGURE 5.1 Typical model of seismic resilience.

of earthquakes (Yodo and Wang 2016). The system functionality, $Q(t)$, is usually a dimensionless percentile and is defined as current functionality normalized by the available functionality prior to the earthquake event. When a hospital is affected by a severe earthquake, it suffers from structural and nonstructural damages, and its functionality is compromised immediately after the earthquake. As effort is directed toward the recovery, the functionality will be recovered. During the repair period, denoted as the recovery time T_{RE}, the functionality stabilizes and might be higher, equal, or less than the functionality prior to the earthquake, depending on the system damage and retrofitting strategies (Bruneau et al. 2003; Miles and Chang 2006). The curve can be updated to consider the damages caused by the earthquake. Even though the shape of the curve describing the recovery progress depends on different manpower and material resources, the method to quantify the system resilience, R_{Resi}, is often similar as in Equation (5.1), where t_{OE} is the time of the earthquake and T_{LC} is the control time of the system. It is noteworthy that the recovery time T_{RE} is used as control time in this study.

$$R_{Resi} = \frac{1}{T_{LC}} \int_{t_{OE}}^{t_{OE}+T_{LC}} Q(t)dt \tag{5.1}$$

An assessment of the seismic resilience of a complicated engineering system allows a better understanding of the impact of earthquakes and can help decision-makers to formulate effective strategies in all phases of the earthquakes (Cimellaro et al. 2010a, b; Alshehri et al. 2015a). To assess the seismic resilience of an engineering system is conceptually straightforward but quantifiably difficult. Several frameworks have been proposed recently. Bruneau et al. (2003) developed a conceptual assessment framework and a system diagram to improve system resilience by system assessment and modification during the pre- and post-earthquake periods. Chang and Shinozuka (2004) proposed a quantification framework that relates the expected losses in future disasters to a community's seismic performance. Miles and Chang (2006) presented a comprehensive conceptual model of functionality recovery that compares the disparity between systems with different levels of disaster preparedness and mitigation decisions. Cimellaro et al. (2016a) proposed a framework named

PEOPLES for measuring community resilience at different spatial and temporal scales using a layered approach. Burton et al. (2017) assessed the impact of an earthquake on community resilience based on the immediate and cumulative loss of permanent housing occupancy, as well as the time to reach a predefined housing capacity. For healthcare systems, there are also some attempts to quantify seismic resilience. Bruneau and Reinhorn (2007) used the percentage of the healthy population, the patients/day treatment capacity, and the repair cost to quantify the resilience and consider the time variation of the functionality. Cimellaro et al. (2010a, b) proposed a framework that integrates loss and recovery organizational efficiency model to quantify the resilience of critical care facilities. A typical hospital building and a hospital network were selected to show the implementation of the proposed framework. Cimellaro et al. (2011) developed a meta-model to estimate the hospital capacity and the dynamic response in real time. The calibrated waiting time is used to evaluate the disaster resilience of healthcare facilities in the proposed meta-model. Zhong et al. (2015) developed a comprehensive framework to identify key indicators to define hospital resilience. The results indicated that the equipment for the on-site rescue was the most important component. Hassan and Mahmoud (2018) used losses resulting from structural damage and lifeline damage to estimate the earthquake-induced functionality reduction of a hospital. The interdependence between hospital and lifeline system is investigated based on the fault tree method (FTA). Markov chain process was then adopted in their following study (Hassan and Mahmoud 2019) to estimate the recovery of the hospital. Recently, Yu et al. (2019) proposed a framework to evaluate the seismic resilience of hospitals based on FTA. The interdependency between the damage of nonstructural components and external supplies of gas and electricity was considered in their framework.

Measuring hospital resilience remains a real challenge, and much research effort is needed, particularly the quantification of the functionality and recovery process. In this study, we propose a framework to quantify the seismic resilience of hospital systems. The proposed framework considers the relative importance of the different components of hospital systems that are determined by expert inquiry. An idealized repair path of the hospital system is incorporated into the framework. The detailed contribution of each recovery sequence which contains the recovery of several different components to the functionality of the system is identified based on the repair path. In addition, the resilience demand on the recovery time of hospital systems is evaluated based on questionnaire survey results. Finally, a case study of a hospital building is conducted to demonstrate the effectiveness of the proposed framework.

5.2 A QUANTITATIVE FRAMEWORK FOR SEISMIC RESILIENCE ASSESSMENT

The proposed framework to quantify the seismic resilience of buildings is illustrated in Figure 5.2. The framework consists of four steps, i.e., the seismic hazard analysis, the fragility analysis, the seismic risk analysis, and the calculation of the seismic resilience. The first three steps are very similar to those used to evaluate the seismic risk of an engineering system. In the first step, a probabilistic seismic hazard analysis (PSHA) is conducted by using a source model and an attenuation model suitable for

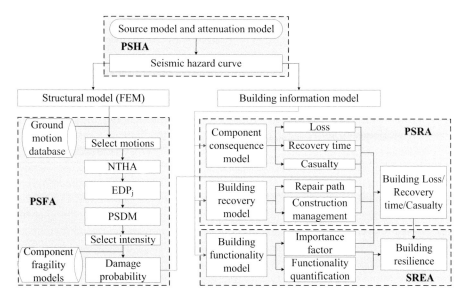

FIGURE 5.2 Quantitative framework for seismic resilience assessment.

the specific seismic environment of the building of interest. The results of the PSHA characterize the joint occurrence of the shaking intensity considering the uncertainties regarding the location, size, mechanism, and magnitude of possible future earthquakes (Bradley et al. 2007; Baker 2008).

The probabilistic seismic fragility analysis (PSFA) is then conducted in the second step. It primarily relies on the finite element model and nonlinear time history analysis (NTHA). The ground motions shall satisfy the characteristics defined by the PSHA and are selected from the strong motion database. Once the NTHA is completed, the engineering demand parameters (EDPs) (e.g., story drift and floor acceleration) are obtained and are used to develop probabilistic seismic demand models (PSDMs). The fragility curves of the different components are obtained based on the PSDMs, and the probability of exceeding different damage states (DS) is also calculated for a selected earthquake intensity (Porter et al. 2007).

A building information model that describes the distribution of the structural components, nonstructural components, equipment, special functional rooms, etc. is necessary for evaluating the seismic effects on a hospital. The building information model contains all components that support the functionality of the building. The component consequence model is composed of the loss model, the recovery time model, and the casualty model. They provide the economic loss, the recovery time, and the number of injured or dead, respectively, caused by a damaged component at a given damage state. Most risk assessment frameworks use the three models. However, they rarely consider the repair process, which determines the repair sequence and requires additional specific consideration. To this end, the building recovery model is proposed and is incorporated into the building information model to take the repair path into account by using the construction management and the importance factor. In this step, a probabilistic seismic risk analysis (PSRA) is conducted to estimate the consequences of a component's damage as a result of an earthquake. The building loss, recovery time, and casualties can be evaluated by integrating the contribution of these components.

The final step is to assess the building resilience using the seismic resilience evaluation analysis (SREA). Various indices have been proposed in past studies to define the functionality of the hospital system. The waiting time of patients, number of beds, number of operating rooms, resources, eventual loss of healthy population, and staff productivity in the hospital were integrated together by Cimellaro et al. (2011) to describe the performance of the hospitals. Paul et al. (2006) adopted waiting time as the main parameter of response to evaluate the earthquake resilience of healthcare facilities, and later, this index was also used in other studies (Welch et al. 2011; Cimellaro and Pique 2015, 2016; Dong and Frangopol 2016; Cimellaro et al. 2016b; Favier et al. 2019). Lupoi et al. (2012) suggested to use the number of functioning operating beds available after an earthquake event as the performance index. However, it is difficult to get such data directly from the Chinese hospitals because of confidential issues. As a compromise, the economic index would be suitable to define the functionality loss and could be acceptable for the governor to count the loss. Therefore, it is very important to quantify the importance factor of each component. In doing so, the seismic resilience can be explicitly quantified by the loss and recovery time. The casualty index is also important and can be incorporated into the building resilience model. However, more effort is needed to accurately define the casualty model, and this is beyond the scope of this study.

In this study, the hospital building information model, the importance factors, the repair path, and the evaluation method of building loss and recovery time are discussed in the following sections.

5.3 HOSPITAL SYSTEM MODEL

A typical hospital system has five functional units, i.e., the emergency department (ED), clinical department (CD), medical and technical department (MTD), medical management department (MMD), and prevention and healthcare department (PHD), as shown in Figure 5.3a (GB 51039-2014 2014). These functional units are supported by seven subsystems containing the structure system, the electrical system, the HVAC system, the medical system, the enclosure system, the water supply and drainage system, and the egress system (Figure 5.3b). Among them, the medical system is unique. Each subsystem is composed of different components as presented in Figure 5.3c.

The functionality of a hospital system following an earthquake mostly depends on the damage of these structural and nonstructural components including the medical equipment. Determining the relative significance of the different components is necessary to calculate the seismic resilience of a hospital system. It is of utmost urgency to recover or repair the most important components that may have a direct influence on the emerging care of injured people during the earthquake. Therefore, the critical functional units, subsystems, and components need to be identified to evaluate seismic resilience systematically.

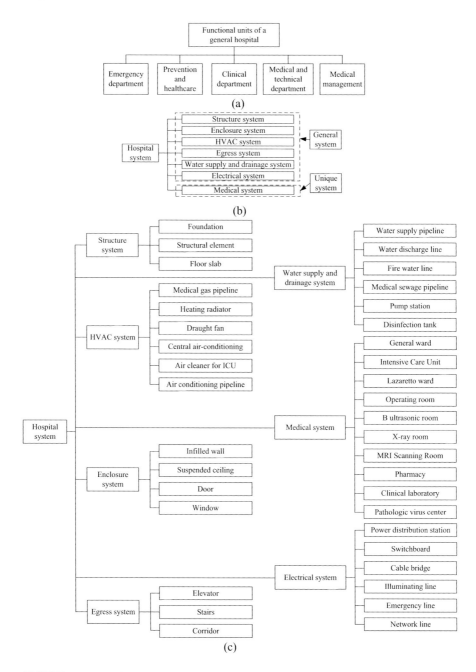

FIGURE 5.3   Hospital system model. (a) Functional units, (b) subsystems, and (c) subsystems and components.

5.4 DETERMINATION OF THE IMPORTANCE FACTORS

The determination of the importance factors of the components requires not only engineering experience but also medical knowledge. Therefore, a panel of 70 experts (Figure 5.4) comprised the expert group that was invited to participate in a questionnaire survey to determine the importance factors. Seventy is an acceptable number when conducting a questionnaire survey (Lee and Walsh 2011). Some experts with professional experience in post-earthquake rescue were involved in the survey group. The experts were recruited from a variety of disciplines, and the respective categories include doctors (DOC), medical facility (MF), earthquake relief work (ER), structural seismic resistance design (SSR), disaster prevention and reduction (DPR), and urban planning (UP). The experts were asked to complete two series of questionnaires. One was the Delphi-based method (DBM), and the other was the analytic hierarchy process (AHP). The importance factors derived from both methods were compared and calibrated.

5.4.1 DELPHI-BASED METHOD

The DBM was developed in the 1950s and is an important systematic data collection method for gathering information and judgment from experts. It includes a group of experts involved in a process to obtain the most reliable consensus on a specific topic (Okoli and Pawlowski 2004). This method has been widely used in various fields including disaster research (Jordan and Javernick-Will 2013; Alshehri et al. 2015b). Once the criteria, scale, and format of the questionnaire are established, a pilot survey is conducted using five participants to examine whether the survey is easy to complete. After the pilot survey, modifications are made to improve its effectiveness. Thereafter, the formal survey is conducted. Feedback and iteration might be necessary to obtain a reliable consensus. The experts are allowed to revise their judgments and change their answers. Figure 5.5 shows the flowchart of the DBM (Alshehri et al. 2015b). In this study, three rounds of survey iterations were conducted.

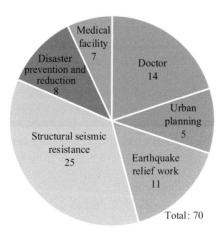

FIGURE 5.4 Experts from different disciplines.

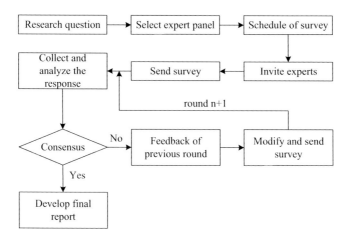

FIGURE 5.5   Flowchart of the DBM.

For the subsystem i composed of n components, each expert needs to give a grade $\sigma_{i,j}$ to the component j on the condition that the total grade of all components in the subsystem is $100 \left(\sum\limits_{j=1}^{n} \sigma_{i,j} = 100 \right)$. Then, the component importance factor, $\omega_{i,j}$, for this component is calculated using Equation (5.2). Similar calculations are conducted for the five functional units and seven subsystems to determine their importance factors.

$$\omega_{i,j} = \frac{\sigma_{i,j}}{\sum\limits_{j=1}^{n} \sigma_{i,j}} \tag{5.2}$$

5.4.2   ANALYTIC HIERARCHY PROCESS

The AHP is another method for determining the importance factors. The AHP has been widely used for determining the relative importance among a series of indicators using pairwise comparisons (Orencio and Fujii 2013; Panahi et al. 2014). The AHP model of a hospital system is presented as a three-tier hierarchy as shown in Figure 5.3c, wherein the top tier represents a disaster-resilient hospital system. The second tier describes the seven subsystems, and the bottom tier represents the components of each subsystem. Based on the AHP model, pairwise comparisons are used to identify the relative importance of the criteria and components associated with a disaster-resilient hospital system. In this study, there were 21 subsystem comparisons involved in a matrix for the seven criteria, whereas the component comparisons for each criterion ranged from 3 to 45. Similar procedures could be applied to the five functional units. Flowchart of the AHP is depicted in Figure 5.6. Each pairwise comparison result is considered as the participant's preference for one indicator over another and is expressed as a set of fundamental scales ranging from 1 to 9, as shown in Table 5.1 (Saaty 1990). a_{ij} is the score of the indicator i over the indicator j based

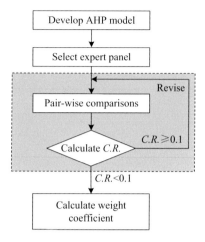

FIGURE 5.6 Flowchart of the AHP.

TABLE 5.1
Rating Scale for Judging Preferences

Scale	Judgment of Preference	Description
1	Equally important	Two factors contribute equally to the objective
3	Moderately important	Experience and judgment slightly favor one over the other
5	Strongly important	Experience and judgment strongly favor one over the other
7	Very strongly important	Experience and judgment very strongly favor one over the other
9	Extremely important	The evidence favoring one over the other is of the highest possible validity
2/4/6/8	Intermediate preferences between adjacent scales	When compromise is needed

on a priority judgment and follows the rule presented in Equations (5.3) and (5.4). The distribution of these scores is described as the square matrix A (Equation 5.5) and is called the consistent matrix.

$$a_{ij} = \frac{1}{a_{ji}}, a_{ij} > 0 \tag{5.3}$$

$$a_{ij} = 1, i = j \tag{5.4}$$

$$A = \left(a_{ij}\right) = \begin{bmatrix} a_{11} & a_{12} & \cdots & a_{1n} \\ a_{21} & a_{22} & \cdots & a_{2n} \\ \cdots & \cdots & \cdots & \cdots \\ a_{n1} & a_{n2} & \cdots & a_{nn} \end{bmatrix} \tag{5.5}$$

TABLE 5.2

The Order of the Random Index of Consistency with a Number of Alternatives

n	1	2	3	4	5	6	7	8	9	10
R.I.	0	0	0.58	0.9	1.12	1.24	1.32	1.41	1.45	1.49

For the AHP, the importance factor (W_i) was obtained using Equations (5.6)–(5.8). The procedure can be conducted several times until the scores are consistent, as shown in Figure 5.6. The consensus of the scores of each pairwise comparison must be achieved before determining the importance factor. The consistency index CI is computed by Equation (5.9), where λ_{max} is the maximum eigenvalue of the matrix A, and n is the number of indicators in the priority judgment. The computed CI is then compared with a random consistency index RI given in Table 5.2 to determine the consistency ratio CR, which determines whether the experts' judgment scores or weights can be accepted. The results are acceptable if CR ≤ 0.10 (Liedtka 2005).

$$M_i = \prod_{j=1}^{n} a_{ij}, i = 1, 2, \ldots, n \tag{5.6}$$

$$w_i^{'} = \sqrt[n]{M_i}, i = 1, 2, \ldots, n \tag{5.7}$$

$$W_i = \frac{w_i^{'}}{\sum_{j=1}^{n} w_j^{'}} \tag{5.8}$$

$$CI = \frac{\lambda_{max} - n}{n - 1} \tag{5.9}$$

$$CR = \frac{CI}{RI} \tag{5.10}$$

5.5 DISCUSSION OF IMPORTANCE FACTORS

The importance factors of the functional units calculated by the two methods are compared in Figure 5.7a. They are generally in agreement; therefore, it is concluded that both methods are reliable and suitable. The importance factors obtained by the AHP are used for the discussion in the following section. With regard to the relative importance of the different functional units, the six expert groups all consider that the ED and the CD are the two most important functional units. The importance factors of the ED are much larger than that of the other units. The values of the importance factors are 0.300, 0.463, 0.389, 0.372, 0.378, and 0.449 for the expert groups

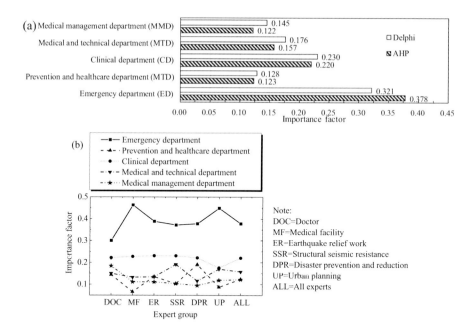

FIGURE 5.7 Importance factors of the functional units. (a) Comparison of the two different methods and (b) results of different expert groups using the AHP method.

of DOC, MF, ER, SSR, DPR, and UP, respectively, as shown in Figure 5.7b. The ED has the highest overall weight and accounts for 37.8% of the hierarchy's total weight. This is consistent with a previous study that emphasized the importance of fulfilling medical first aid and health service rescue to help the wounded after earthquakes (Cimellaro et al. 2011). The other functional units have the following importance factors: CD (22.0%), MTD (15.7%), PHD (12.3%), and MMD (12.2%).

Table 5.3 presents the importance factors of the seven subsystems, among which the structure system is the most important one, because the structurally damaged hospital building is not allowed to provide service after an earthquake, meaning a complete loss of functionality. The medical system directly related to the hospital functionality is the second important subsystem. Electrical system, and water supply and drainage system are lifelines, and the importance factors of these two subsystems are comparable to those of structure system and medical system. Table 5.4 presents detailed information on the importance factors of the components in each subsystem. It was found that the operating room (17.8%), the intensive care unit (14.3%), the general ward (11.4%), and the lazaretto ward (11.3%) are the four most important components of the medical system. Figure 5.8 lists the importance factors $\omega_{(i,j)}$ of the components considering the importance of the subsystems; the data represent the global component importance factor identified by the AHP method. It is also interesting that the foundation is of the highest importance, and this may be attributed to the universal knowledge that once the foundation fails, the entire building cannot provide any services.

TABLE 5.3
Importance Factors of the Subsystems (AHP)

Subsystem	Doctor	Medical Facility	Earthquake Relief Work	Structural Seismic Resistance	Disaster Prevention and Reduction	Urban Planning	All
Structure system	0.198	0.275	0.300	0.271	0.343	0.296	0.274
Water supply and drainage system	0.141	0.117	0.142	0.100	0.104	0.156	0.122
HVAC system	0.115	0.108	0.072	0.095	0.062	0.099	0.093
Electrical system	0.140	0.150	0.126	0.149	0.174	0.093	0.142
Enclosure system	0.104	0.071	0.062	0.089	0.069	0.122	0.087
Medical system	0.183	0.193	0.174	0.209	0.142	0.148	0.181
Egress system	0.119	0.086	0.124	0.087	0.106	0.085	0.102

TABLE 5.4
Importance Factors of the Components in Each Subsystem (AHP)

Subsystem	ω_i^a	Component	$\omega_{i,j}^b$	$\omega_{(i,j)}^c$
Structure system	0.274	Foundation	0.459	0.126
		Structural element	0.376	0.103
		Floor slab	0.165	0.045
Water supply and drainage system	0.122	Water supply pipeline	0.208	0.025
		Water discharge line	0.153	0.019
		Firewater line	0.202	0.025
		Medical sewage pipeline	0.152	0.018
		Pump station	0.160	0.019
		Disinfection tank	0.126	0.015
HVAC system	0.093	Medical gas pipeline	0.285	0.027
		Heating radiator	0.137	0.013
		Draught fan	0.121	0.011
		Central air conditioning	0.120	0.011
		Air cleaner for ICU	0.236	0.022
		Air conditioning pipeline	0.101	0.009
Electrical system	0.142	Power distribution station	0.295	0.042
		Switchboard	0.159	0.023
		Cable bridge	0.115	0.016
		Illuminating line	0.134	0.019
		Emergency line	0.205	0.029
		Network line	0.092	0.013
Enclosure system	0.087	Infilled wall	0.378	0.033
		Suspended ceiling	0.244	0.021
		Door	0.215	0.019
		Window	0.163	0.014

(*Continued*)

TABLE 5.4 (*Continued*)
Importance Factors of the Components in Each Subsystem (AHP)

Subsystem	ω_i[a]	Component	$\omega_{i,j}$[b]	$\omega_{(i,j)}$[c]
Medical system	0.181	General ward	0.114	0.021
		Intensive care unit	0.143	0.026
		Lazaretto ward	0.113	0.020
		Operating room	0.178	0.032
		B ultrasonic room	0.072	0.013
		X-ray room	0.081	0.015
		MRI scanning room	0.070	0.013
		Pharmacy	0.086	0.016
		Clinical laboratory	0.074	0.013
		Pathologic virus center	0.068	0.012
Egress system	0.102	Elevator	0.355	0.036
		Stairs	0.440	0.045
		Corridor	0.206	0.021
Sum	1.000	-	-	1.000

[a] ω_i is the importance factor of the subsystem *i*.
[b] $\omega_{i,j}$ is the importance factor of the component *j* of the subsystem *i*.
[c] $\omega_{(i,j)}$ is the importance factor of the component *j* considering the importance of the subsystems *i*, and
$\omega_{(i,j)} = \omega_i \cdot \omega_{i,j}$.

5.6   DEMAND ON RECOVERY TIME

Along with the questionnaire on the importance factors, the desirable recovery time is also investigated by asking the participants to provide their general opinion from the perspective of each discipline. The desirable recovery time is the maximum acceptable period for a hospital to provide no services. In this study, the desirable recovery times for the emergency functionality and the complete functionality were both evaluated based on the five functional units. The results are presented in Table 5.5. The emergency functionality is most important during the 72-hour "golden window" for survivors, whereas the recovery time of the complete functionality may be longer. Note that the total recovery times in Table 5.5 can be considered the upper limit when the recovery process is conducted in sequence. The emergency department shall be repaired first because it has the largest importance factor. It is estimated that 97% of earthquake-related injuries occur immediately or within the first 30 minutes after the main shock (Gunn 1995). Emergency medical care is required as soon as possible. For the emergency functionality, the recovery of the emergency department is most urgent and should occur within 0.1 days (2.4 hours). However, the mean recovery times for emergency departments of local hospitals after the 1999 Kocaeli (Turkey) and 1999 Chi-Chi (Taiwan) earthquakes were 11.86 days (284.6 hours) and 0.49 days (13.3 hours) according to the data reported in (Myrtle et al. 2005). None of them meet the requirement of emergency rescue.

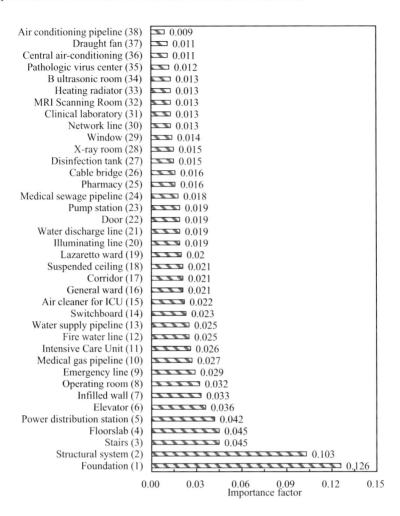

FIGURE 5.8 Importance factors of the components considering the importance of the subsystems.

TABLE 5.5
Desirable Recovery Time of the Hospital System (unit: day)

	Desirable Recovery Time	
Functional Units	Emergency Functionality Recovery	Complete Functionality Recovery
Emergency department	0.1	1.45
Prevention and healthcare department	1.1	5.24
Clinical department	0.93	4.25
Medical and technical department	0.39	5.4
Medical management department	5.63	10.93
Total	8.13	27.27

5.7 ASSESSMENT OF SEISMIC RESILIENCE

The seismic resilience of an engineering system is often assessed using three indices, i.e., the recovery time, the economic loss, and the casualties (Cimellaro et al. 2010a, 2010b). In this study, the economic loss caused by the related components is used to quantify the corresponding functionality. This is a compromise because of the lack of related data on functionality-related indicators (such as the waiting time of patients and number of medical staffs). However, this is reasonable because any functionality is supported by components, and the damage or failure of any of them would compromise the functionality. The damage or failure of components can be quantified by the induced economic loss. Therefore, the economic loss of components is adopted to evaluate the functionality. The casualty index is also important. However, it depends on so many parameters that it requires more effort for an accurate calculation. In this study, the casualty index is not considered in the seismic resilience assessment.

5.7.1 RECOVERY TIME OF COMPONENTS

The recovery time of a component is expressed in Equation (5.11), where N is the total number of damage states, rt_i is the median recovery time associated with the damage state i and $P_{DS_i|IM}$ is the conditional probability of the component in the damage state i (DS_i) after an earthquake with a certain intensity measure [IM, e.g., PGA, S_a (T_1, 5%)]. The probabilities of exceedance given a specific damage state can be obtained by the PSFA, and $P_{DS_i|IM}$ is then calculated as the difference between the probabilities of exceeding the damage states i and $i+1$, where the damage state $i+1$ is more severe than the damage state i (Porter et al. 2007).

$$rt = \sum_{i=1}^{N} rt_i \cdot P_{DS_i|IM} \tag{5.11}$$

5.7.2 ECONOMIC LOSS OF COMPONENTS

A dimensionless factor, the repair cost ratio, is used to define the economic loss in this study. The repair cost is defined as the direct cost of the material, construction machines, and human resources spent in recovering a damaged component to its original state prior to the earthquake. It is often scaled by the direct cost to completely re-construct one identical component, which is called the repair cost ratio or the economic loss ratio. This is expressed similarly as the recovery time and is defined in Equation (5.12) where L_i is the median economic loss ratio associated with the damage state i. The economic loss ratio of a component ranges from 0 to 1.

$$L = \sum_{i=1}^{N} L_i \cdot P_{DS_i|IM} \tag{5.12}$$

5.7.3 Repair Path

The repair cost of a building depends not only on the component repair cost but also on the repair path, which is largely determined by the location and construction sequence. A damaged ceiling, for example, cannot be repaired before the restoration of the structural element, whereas all ceilings can be restored simultaneously if sufficient resources are available. In this study, the repair of components will take most of the time, and the repair time is adopted as an approximation of the time to resume functions. In the proposed framework, the repair time is calculated by using a repair path in the similar way as the platforms such as HAZUS-MH MR3 (2003), PEER framework (Cornell and Krawinkler 2000), and ATC-58 (2009). But the repair path used in this study is more realistic than that used in ATC-58 where the components are fixed either in a completely parallel way or in a series way. The repair path used in this study combines the two patterns considering the actual repair logic. As shown in Figure 5.9, the idealized repair path has seven steps (or sequences) that occur in series, whereas in each step, several component repairs can proceed in parallel. For one parallel sequence, the recovery time is determined by the component that takes the longest time to restore. Once the recovery time of each step is obtained, the recovery time of the building is calculated as the sum of the recovery times of all seven steps. It is also assumed that the recovery work would be conducted in all stories simultaneously, and the recovery time of one specific component is the maximum time needed for repairing that component in each story.

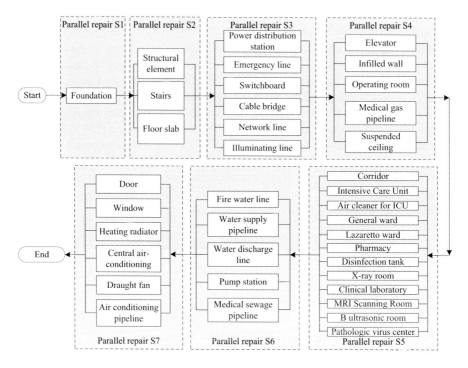

FIGURE 5.9   Idealized repair path of the hospital system.

However, it is worth noting that the real repair path should be determined by related experts, including structure engineers, system engineers, doctors, and hospital managers. And this real repair path can be easily incorporated into the proposed framework. Note that the financial resources, availability of contractors, the preparation of specialized equipment, and any other factors that may extend the re-occupancy period of the building are not considered here.

5.7.4 RESILIENCE ASSESSMENT

The 38 components listed in Figure 5.8 are used with the component importance factor represented by ω_i and the functionality loss of $L_{(i)}$ for component i. L_i^* is the weighted loss of the functionality and is defined in Equation (5.13). The total functionality loss of the hospital system, *Loss*, is calculated as the sum of the weighted losses of all components, as defined in Equation (5.14). Subsequently, the functionality of the hospital system at time t_{OE} ($Q(t_{OE})$) can be calculated as shown in Equation (5.15a), the functionality after recovery sequence i ($Q(t_{S_i})$) is calculated by Equation (5.15b), where t_{S_i} represents the time when recovery sequence i is completed, $\sum_j L_{ij}$ represents the functionality loss (or functionality improvement after recovery) caused by the components belonging to recovery sequence i and is calculated as the sum of the weighted losses (see Equation 5.13) of these components. RT is the generic sum of the recovery time of the 38 components and is defined in Equation (5.16a), where the symbol $\sum^*$ represents the generic sum considering the idealized repair path described previously. The recovery duration time of recovery sequence i (RT_{S_i}) is calculated as Equation (5.16b), and t_{S_i} is obtained by Equation (5.16c), where $\max_j \{rt_{ij}\}$ is the maximum time needed to repair the components belonging to recovery sequence i. The schematic to calculate the seismic resilience is presented in Figure 5.10, where the functionality is given by Equation (5.15) and the recovery time by Equation (5.16). It should be noted that the curve between the initial point and the end point of the recovery process is defined as the recovery function, which reflects the redundancy of the system and the resourcefulness to be cast into the repairing (Lin et al. 2016). Simplified recovery functions as shown in Figure 5.11 have been proposed in previous studies (Cimellaro et al. 2010a, b; Zhao et al. 2017). They are usually employed in the resilience analysis and are referred to as the linear recovery, the exponential recovery, the trigonometric recovery, and the non-recovery, respectively, where a and b are constant values.

$$L_i^* = \omega_i L_{(i)} \tag{5.13}$$

$$\mathrm{Loss} = \sum_{i=1}^{38} L_i^* = \sum_{i=1}^{38} \omega_i L_{(i)} \tag{5.14}$$

$$Q(t_{OE}) = 1 - \mathrm{Loss} = 1 - \sum_{i=1}^{38} \omega_i L_{(i)} \tag{5.15a}$$

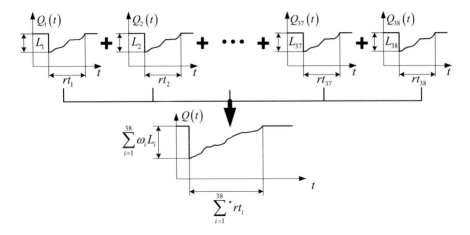

FIGURE 5.10 Schematic of the seismic resilience calculation.

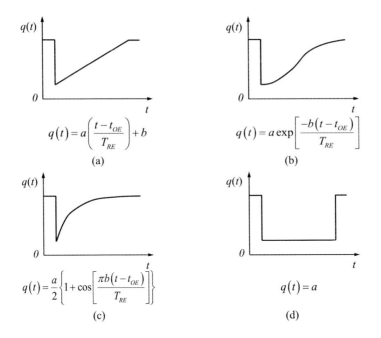

FIGURE 5.11 Simplified recovery functions. (a) Linear recovery, (b) trigonometric recovery, (c) exponential recovery, and (d) non-recovery.

$$Q(t_{S_i}) = \begin{cases} Q(t_{OE}) + \sum\limits_{j} L_{ij} & i = 1 \\ Q(t_{S_{i-1}}) + \sum\limits_{j} L_{ij} & i = 2,3,\cdots,7 \end{cases} \tag{5.15b}$$

$$RT = \sum_{i=1}^{38} {}^* rt_i \tag{5.16a}$$

$$RT_{S_i} = \max_{j} \left\{ rt_{ij} \right\} \tag{5.16b}$$

$$t_{S_i} = \begin{cases} t_{OE} + RT_{S_i} & i = 1 \\ t_{S_{i-1}} + RT_{S_i} & i = 2, 3, \cdots, 7 \end{cases} \tag{5.16c}$$

The final resilience curve describes the functionality of a hospital system following an earthquake, based on which the resilience can be quantified using Equation (5.1). During the entire procedure of quantifying the resilience of an engineering system, the assumptions have a strong effect on the accuracy of the assessment results (Dong et al. 2013; 2014). It is wise to consider the real conditions of a system including the financial resources, availability of contractors, and so on. Although thorough studies are necessary to develop sophisticated models of such factors, they can be easily implemented in the proposed framework.

5.8 CASE STUDY OF A HOSPITAL BUILDING

5.8.1 INTRODUCTION TO THE HOSPITAL BUILDING

The hospital building used for the case study was a 10-story reinforced concrete (RC) frame (Figure 5.12). Table 5.6 presents the story heights and story mass of the building. The uniformly distributed dead and live loads on each floor were 3.0 and 2.0 kN/m², respectively, and those for the roof were 4.0 and 0.5 kN/m², respectively. It was designed following the Chinese Code for Seismic Design of Buildings (GB 50011-2010 2010), and the PGA of the design-basis earthquake ground motion was 0.2 g. The site was classified as type II, and the design characteristic period was 0.35 s. The layout of the building was significantly irregular. The concrete was C30 with a standard compressive strength of 20.1 Mpa, and the steel rebars were HRB400 with a yielding strength of 360 MPa.

A finite element model of the hospital building was created in the OpenSees software. The beams and columns were simulated using nonlinear beam-column elements, which were defined as force-based elements with distributed plasticity. Five Gauss-Lobatto integration points were inserted along the length of each element with two at the ends to simulate the formation of plastic hinges. The composite RC cross-sections were conveniently simulated by the fiber formulation. The concrete material is represented by Hisham's model (Hisham and Yassin 1994) with a linearly decaying tensile strength. The input parameters of the confined concrete were determined using Mander's theoretical stress-strain model (Mander et al. 1988). A Giuffre-Menegotto-Pinto uniaxial material model with isotropic strain hardening was used to model the longitudinal steel rebars. In the finite element model, the story mass and the corresponding gravity were uniformly distributed to each beam-column joint. It was assumed that the foundation of the hospital will not be damaged after earthquakes, and thus, the foundation was not involved in the finite element model.

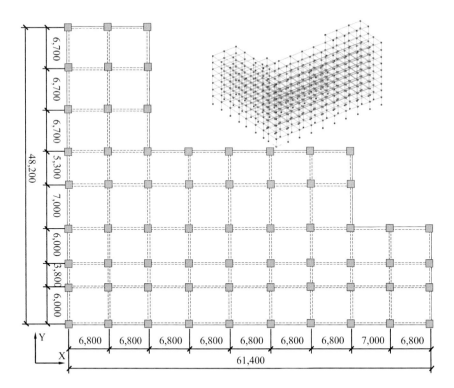

FIGURE 5.12   Plan view and three-dimensional model of the hospital building (unit: mm).

TABLE 5.6
Story Height and Mass of the Hospital Building
(units: m, ton)

Story Number	Story Height	Total Story Mass
10	3.90	224.9
9	3.45	1108.8
8	3.45	1292.6
7	3.45	1991.8
6	3.45	1980.8
5	3.45	1980.8
4	3.45	1980.8
3	3.45	2213.0
2	3.95	2492.1
1	3.80	2532.6

The structure was assumed to be fixed to the ground and the rigid diaphragm constraint was used for each floor. A geometric nonlinearity was considered for the
P-Delta effect. A Rayleigh damping of 5% was assigned to the first two modes of
vibration. The first three fundamental natural periods of the building were 1.2362,
1.1599, and 1.0772 s, respectively.

5.8.2 SELECTED GROUND MOTIONS

Without the background information regarding earthquakes and geologic information on the area where the building is located, the PSHA cannot be conducted.
Instead, the far-field record set that includes 22 records suggested by FEMA P695
(2009) was adopted to evaluate the seismic performance of the hospital building.
These 22 records were obtained from 14 events that occurred between 1971 and
1999. More information about the ground motions can be found in Appendix A.9
of FEMA P695 (2009). The acceleration response spectra of the selected ground
motions are presented in Figure 5.13, assuming the damping ratio was 5%. The thick
line is the mean spectrum curve.

5.8.3 SEISMIC FRAGILITY ANALYSIS

The probabilistic seismic fragility analysis (PSFA) was conducted based on the
results of nonlinear time history analysis (NTHA) and predefined damage states
or performance levels described by component fragility models. Responses of the
structure including maximum interstory drift ratio (IDR) and peak floor acceleration (PFA) of each story to a series of ground motions (Figure 5.13) were obtained
by NTHA. Then, a probabilistic seismic demand model (PSDM) was obtained
through regression analysis using the computed responses, as the example shown in
Figure 5.14. A PSDM describes the relationship between the engineering demand

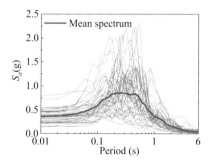

FIGURE 5.13   Response spectra of the selected ground motions.

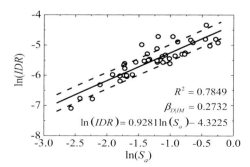

FIGURE 5.14   Example of probabilistic seismic demand model.

parameter (EDP, such as IDR and PFA) of the considered component and ground motion intensity measure [IM, such as PGA or $S_a\,(T_1, 5\%)$]. In this study, PSDM was estimated using a standard power function (Cornell et al. 2002):

$$S_D = a(IM)^b \tag{5.17}$$

which can be rewritten as Equation (5.18) in the logarithmically transformed space after taking logarithms:

$$\ln(S_D) = \ln(a) + b\ln(IM) \tag{5.18}$$

where S_D is the conditional median of the EDP given the IM, and a and b are the parameters of the regression which can be obtained from linear regression in the logarithmically transformed space. In addition to the median values, uncertainty associated with the demand model was characterized by a lognormal distribution in which the conditional logarithmic standard deviation, $\beta_{D|IM}$, was estimated based on the demand data:

$$\beta_{D|IM} \cong \sqrt{\frac{\sum_{i=1}^{M}\left(\ln(d_i) - \ln\left(a(IM)^b\right)\right)^2}{M - 2}} \tag{5.19}$$

where M is the number of numerical simulation analyses, and d_i is the calculated peak demand quantity of i-th numerical simulation analysis. Figure 5.14 shows an example of the results of the probabilistic seismic demand analysis (PSDA), where R^2 is the coefficient of determination indicating the robustness of the regression. Parameters used to define IDR and PFA of each story were obtained from PSDA and shown in Table 5.7. For the maximum IDR, $S_a\,(T_1, 5\%)$, as suggested by FEMA-P58 (2012) was used as the intensity measure and PGA was used for the PFA.

The seismic fragility can be simply defined as the conditional probability that the seismic demand (D) placed on the component exceeds its capacity (C) for a given

TABLE 5.7
Parameters of the Probabilistic Seismic Demand Model

Number of Story	Maximum Interstory Drift Ratio				Peak Floor Acceleration			
	b	$\ln(a)$	R^2	$\beta_{D|IM}$	b	$\ln(a)$	R^2	$\beta_{D|IM}$
1	0.9281	−4.3225	0.7849	0.2732	0.9383	−0.6160	0.7338	0.2360
2	0.8324	−3.9517	0.7983	0.2353	0.8893	−0.0366	0.8070	0.1816
3	0.8770	−3.7688	0.8341	0.2200	0.9403	0.2604	0.8753	0.1482
4	0.9129	−3.6973	0.8617	0.2057	1.0084	0.4192	0.9037	0.1374
5	0.8988	−3.7879	0.8644	0.2002	0.9987	0.4145	0.9023	0.1372
6	0.8467	−4.0145	0.8209	0.2224	0.9399	0.3161	0.8591	0.1589
7	0.8241	−4.2328	0.7679	0.2548	0.8851	0.2952	0.8256	0.1699
8	0.9670	−3.7654	0.5318	0.5103	0.7601	0.1988	0.7277	0.1941
9	0.6570	−4.5230	0.4051	0.4478	0.8097	0.4373	0.7991	0.1695
10	0.5797	−4.9178	0.3252	0.4696	0.9143	0.8476	0.8086	0.1857

IM level. With the developed PSDM of the component and the component capacity model which is usually referred to as the component fragility model (Porter et al. 2007), the exceedance probability that the demand would be larger than the capacity was computed as Equation (5.20). It can be rewritten as Equation (5.21) by substituting the demand median S_D in the form of Equation (5.18) (Solberg et al. 2010):

$$P[D \geq C \mid IM] = \Phi\left(\frac{\ln(S_D / S_C)}{\sqrt{\beta_{D|IM}^2 + \beta_C^2}} \right) \tag{5.20}$$

$$P[D \geq C \mid IM] = \Phi\left(\frac{\ln(IM) - (\ln(S_C) - \ln(a))/b}{\sqrt{\beta_{D|IM}^2 + \beta_C^2 / b}} \right) \tag{5.21}$$

where S_C and β_C are median and logarithmic standard deviation used to define the component fragility model. Fragility curve for the considered components in the building can then be obtained using Equation (5.21). The fragility parameters (S_C and β_C), median loss ratio (L_i as defined in Equation 5.12), median recovery time (rt_i as defined in Equation 5.11), and floor distribution of the 38 components used in the resilience assessment are given in Table 5.8.

TABLE 5.8
Components and Damage States

Component	Floor Distribution	E D P	References	DS	Median	Dispersion	Loss Ratio (Median)	Recovery Time (Median)
Foundation	–	– –		–	–	–	–	–
Structural system	1–10	I D R	HAZUS-MH MR3 (2003)	DS_1	0.0025	0.2	0.014	0.52
				DS_2	0.005	0.2	0.1	3.71
				DS_3	0.015	0.3	0.5	18.54
				DS_4	0.04	0.4	1	37.08
Stairs	1–10	I D R	FEMA P58 (2012)	DS_1	0.005	0.6	0.1	1.62
				DS_2	0.017	0.6	0.5	8.10
				DS_3	0.028	0.45	1	16.20
Floor slab	1–10	I D R	FEMA P58 (2012)	DS_1	0.025	0.25	0.5	8.10
				DS_2	0.04	0.25	1	16.20
Power distribution station	–	P G A	FEMA P58 (2012)	DS_1	2.16	0.45	1	15.12
Elevator	1–10	P F A	FEMA P58 (2012)	DS_1	0.5	0.3	1	12.96
Infilled wall	1–10	I D R	Teng et al. (2018)	DS_1	0.0005	0.11	0.014	0.17
				DS_2	0.0013	0.29	0.1	1.19
				DS_3	0.0040	0.55	0.3	3.56
				DS_4	0.0127	0.29	0.5	5.94
				DS_5	0.0246	0.28	1	11.88

(Continued)

TABLE 5.8 (*Continued*)
Components and Damage States

Component	Floor Distribution	EDP	References	DS	Median	Dispersion	Loss Ratio (Median)	Recovery Time (Median)
Operating room	6–7	I D R	Wang (2005)	DS$_1$	0.002	0.4	1	11.52
Emergency line	1–10	P F A	FEMA P58 (2012)	DS$_1$ DS$_2$	1.5 2.25	0.5 0.5	0.7 1	7.31 10.44
Medical gas pipeline	6–7	P F A	FEMA P58 (2012)	DS$_1$ DS$_2$	1.5 2.25	0.5 0.5	0.7 1	6.80 9.72
Intensive care unit	4–5	I D R	Wang (2005)	DS$_1$	0.002	0.4	1	9.36
Firewater line	1–10	I D R	Wang et al. (2019)	DS$_1$ DS$_2$ DS$_3$	0.0566 0.0891 0.1195	0.285 0.2907 0.2686	0.1 0.5 1	0.90 4.50 9.00
Water supply pipeline	1–10	P F A	FEMA P58 (2012)	DS$_1$ DS$_2$	1.5 2.25	0.5 0.5	0.7 1	6.30 9.00
Switchboard	1–10	P F A	FEMA P58 (2012)	DS$_1$	2.16	0.45	1	8.28
Air cleaner for ICU	4–5	P F A	FEMA P58 (2012)	DS$_1$	1.54	0.6	1	7.92
General ward	3–5	I D R	Wang (2005)		0.003	0.4	1	7.56
Corridor	-	I	-	-	-	-	-	-
Suspended ceiling	1–10	P F A	Soroushian et al. (2016)	DS$_1$ DS$_2$ DS$_3$	0.98 1.329 1.791	0.396 0.401 0.419	0.05 0.2 1	0.38 1.51 7.56
Lazaretto ward	3	I D R	Wang (2005)	DS$_1$	0.002	0.4	1	7.20
Illuminating line	1–10	P F A	FEMA P58 (2012)	DS$_1$	2.25	0.5	1	6.84
Water discharge line	1–10	P F A	FEMA P58 (2012)	DS$_1$	2.25	0.5	1	6.84
Door	1–10	I D R	FEMA P58 (2012)	DS$_1$ DS$_2$	0.02 0.05	0.5 0.5	0.2 1	1.37 6.84
Pump station	–	P G A	FEMA P58 (2012)	DS$_1$	2.25	0.5	1	6.84

(Continued)

TABLE 5.8 (*Continued*)
Components and Damage States

Component	Floor Distribution	EDP	References	DS	Median	Dispersion	Loss Ratio (Median)	Recovery Time (Median)
Medical sewage pipeline	6–7	PFA	FEMA P58 (2012)	DS_1	2.25	0.5	1	6.48
Pharmacy	1	IDR	Wang (2005)	DS_1	0.003	0.4	1	5.76
Cable bridge	1–10	PFA	FEMA P58 (2012)	DS_1	2.25	0.5	1	5.76
Disinfection tank	–	PGA	FEMA P58 (2012)	DS_1	2.25	0.5	1	5.40
X-ray room	2	IDR	Wang (2005)	DS_1	0.003	0.4	1	5.40
Window	1–10	IDR	FEMA P58 (2012)	DS_1	0.021	0.45	1	5.04
Network line	1–10	PFA	FEMA P58 (2012)	DS_1	2.25	0.5	1	4.68
Clinical laboratory	2	IDR	Wang (2005)	DS_1	0.003	0.4	1	4.68
MRI scanning room	2	IDR	Wang (2005)	DS_1	0.003	0.4	1	4.68
Heating radiator	1–10	PFA	FEMA P58 (2012)	DS_1	2.25	0.5	1	4.68
B ultrasonic room	2	IDR	Wang (2005)	DS_1	0.003	0.4	1	4.68
Pathologic virus center	6	IDR	Wang (2005)	DS_1	0.002	0.4	1	4.32
Central air conditioning	1–10	PFA	FEMA P58 (2012)	DS_1	1.54	0.6	1	3.96
Draught fan	1–10	PFA	FEMA P58 (2012)	DS_1	1.9	0.4	1	3.96
Air conditioning pipeline	1–10	PFA	FEMA P58 (2012)	DS_1	1.5	0.4	0.078	0.25
				DS_2	2.25	0.4	1	3.24

5.8.4 Seismic Resilience Assessment

A seismic resilience assessment was conducted following the method proposed previously. The recovery time and the economic loss of the 38 components corresponding to the service level earthquake (SLE, PGA = 0.07 g), design-basis earthquake (DBE, PGA = 0.2 g), and maximum considered earthquake (MCE, PGA = 0.4 g) are presented in Figures 5.15 and 5.16, respectively. The operating room, intensive care unit, general ward, Lazaretto ward, X-ray room, clinical laboratory, MRI scanning room, B ultrasonic room, and pathologic virus center almost lost 100% of the functionality at the intensity of the MCE.

The functionality loss of the 38 components considering the components' contributions to the hospital system is illustrated in Figure 5.17. The operating room,

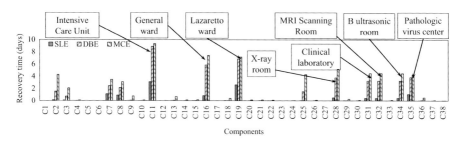

FIGURE 5.15 Recovery time of the 38 components.

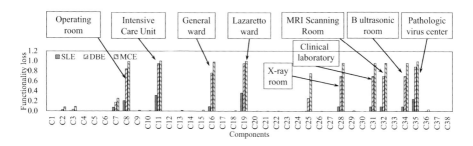

FIGURE 5.16 Functionality loss of the 38 components.

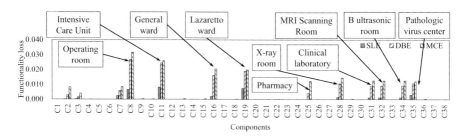

FIGURE 5.17 Functionality loss of the 38 components considering their contribution to the system.

intensive care unit, general ward, Lazaretto ward, pharmacy, X-ray room, clinical laboratory, and MRI scanning room have the largest contributions to the recovery time and the functionality loss.

The recovery time and functionality improvement of the seven recovery sequences are shown in Figure 5.18. It is evident that Sequence 5 is most important for the recovery process. In Sequence 5, the recovery of the intensive care unit, Lazaretto ward, pathologic virus center, general ward, X-ray room, clinical laboratory, MRI scanning room, and B ultrasonic room require most of the time, and the losses of these components represent the largest portion of the functionality loss of the hospital system as reflected in Figures 5.17 and 5.18b. These components all belong to the medical system of the seven subsystems, and this highlights the uniqueness of hospital systems. Considering the importance of the structure system, Sequence 2 also has a long recovery time. In Sequence 4, the Operating room and Infilled wall slow down the recovery process. The identification of these components helps decision-makers to improve the building.

Using the recovery model in Figure 5.11a, the resilience curves for the different earthquake intensities were developed and are presented in Figure 5.19. The resilience values of the different earthquake intensities determined by Equation (5.1) are listed in the last row of Table 5.9. Once an earthquake occurs, the functionality of the hospital is affected and decreases by 3.46%, 14.93%, and 19.73% for the SLE, DBE, and MCE, respectively. The functionality loss was calculated and is based on the economic loss of the five functionalities weighted by the importance factors; the results are shown in Table 5.10 and indicate that the emergency department and clinical department account for most of the losses. The recovery time to achieve complete functionality meets the 27.27-day demand for all three intensities. However, the recovery time of the emergency functionality meets the 8.13-day demand only

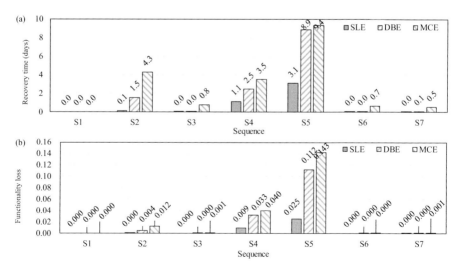

FIGURE 5.18 Recovery time and functionality improvement of the recovery sequences. (a) Recovery time and (b) functionality improvement.

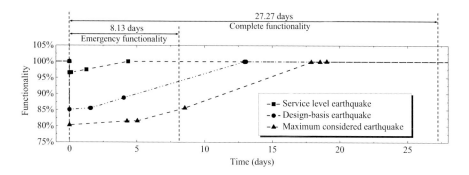

FIGURE 5.19   Resilience curves for the different earthquake intensities.

TABLE 5.9
Resilience of the Hospital System

Earthquake Intensity	SLE	DBE	MCE
PGA (g)	0.07	0.2	0.4
S_a (g)	0.0513	0.1440	0.2883
Recovery time (days)	4.37	13.05	19.05
Functionality loss	3.46%	14.93%	19.73%
R_{Resi}	0.9824	0.9239	0.9074

TABLE 5.10
Functionality Loss (%)

Functional Units	SLE	DBE	MCE
Emergency department	1.31	5.64	7.45
Clinical department	0.76	3.29	4.34
Medical and technical department	0.54	2.35	3.10
Prevention and healthcare department	0.43	1.84	2.43
Medical management department	0.42	1.82	2.40

for the SLE intensity. The desirable recovery times are not achieved for the DBE and MCE intensities. Taking the emergency department as an extreme consideration, it is quite difficult to achieve the recovery in 0.1 days for any of the examined intensities. Therefore, the hospital building in this case study is not robust enough to provide emergency services after an earthquake.

5.8.5 The Effects of Different Recovery Strategies

The repair sequences suggested by REDi (Almufti and Willford 2013), as shown in Figure 5.20, are adopted in this study to demonstrate the effects of different recovery strategies. It is assumed by REDi that the repair of nonstructural components (Sequence A–Sequence G) cannot be started unless the structure is repaired. The structure is repaired one story by one story from the bottom to the top. The downtime due to delays, for instance, the time required to set up the temporary elevator, is not considered here. Sequences A-G are assumed to be repaired simultaneously at one story and in parallel at different floors. The components belonging to each repair sequence are listed in Figure 5.20 with component numbers, e.g., C4 represents floor slabs (see Figure 5.8).

    The resilience curves based on the repair sequence suggested by REDi and the idealized repair path proposed in this study are compared in Figure 5.21. Only the results under MCE are discussed here due to the length of this study. It is obvious that the recovery process can be quite different when different recovery strategies are used. The recovery time to full functionality is 40.87 days following REDi repair sequences, while that is 19.05 days if the proposed idealized repair path is used. The resilience index (R_{Resi}) is 0.8287 for REDi and 0.9074 for the proposed idealized

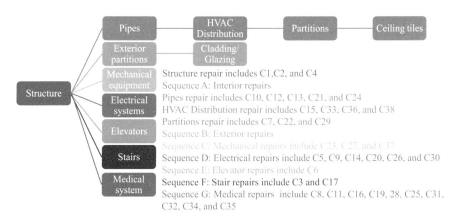

FIGURE 5.20   Repair sequence suggested by REDi (Almufti and Willford 2013).

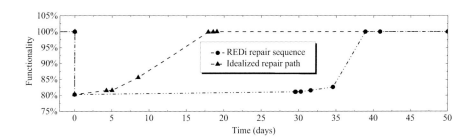

FIGURE 5.21   Resilience curves based on different repair sequences.

repair path. It is noteworthy that both repair sequences are based on some assumptions. The recovery process of a damaged building by earthquake is a complicated issue and still needs further study.

5.9   SUMMARY AND CONCLUSIONS

In this study, we presented a framework for quantifying the seismic resilience of hospital systems so that decision-makers can establish needs and priorities for mitigating the damage of earthquakes. This framework also provides a comprehensive understanding of the damages, responses, and recovery process of the hospital systems. The following conclusions are obtained:

1. A three-tier system model of a hospital system was developed, and the importance factors of the different functional units, subsystems, and components were determined using a questionnaire provided to a multidisciplinary expert panel. It is found that the emergency department has the highest weighting (37.5%) of the five functional units, followed by the structure system (27.4%) and the medical system (18.1%). Among the 38 components, the foundation (12.6%) and the structural element (10.3%) are the most important components.
2. The desirable recovery time of the hospital system was determined based on the questionnaire results and was used to evaluate the hospital resilience. The desirable recovery times of the emergency functionality and the complete functionality are 8.13 and 27.27 days, respectively. The desirable recovery time of the emergency department, which is required to help injured persons immediately after a severe earthquake, is 0.1 day or 2.4 hours.
3. The resilience curve of the hospital system is quantified based on the economic loss weighted by the importance factors and the recovery time is determined based on the repair model. The case study of a typical hospital system indicated that it was not resilient enough to recover the emergency functionality after a severe earthquake.

It is worth noting that the purpose of this study is to develop an assessment framework to quantify the seismic resilience of hospital systems. The accuracy highly depends on the system model and the quality of the available data such as the fragility parameters, the loss ratios, and the recovery model. The parameters used in the case study may not be sufficiently accurate but can be improved by extensive studies.

As a tentative study, this work bears some limitations and several extensions need to be explored. First, the staff behavior (e.g., doctors and nurses) and the injured residents' spatial accessibility to hospital are not included in the resilience assessment framework. Second, the effects of repair crew availability and labor skill are not considered in the repair path and shall be considered in the future study. Third, functionality-related indicators such as waiting time of patients and number of medical staffs should be considered to achieve more comprehensive assessment results.

REFERENCES

Alshehri S A, Rezgui Y, Li H, 2015a. Delphi-based consensus study into a framework of community resilience to disaster, *Natural Hazards*, **75**(3), 2221–2245.

Alshehri S A, Rezgui Y, Li H, 2015b. Disaster community resilience assessment method: A consensus-based Delphi and AHP approach, *Natural Hazards*, **78**(1), 395–416.

ATC-58, 2009. *Guidelines for Seismic Performance Assessment of Buildings.* Applied Technology Council (ATC), Washington, DC

Baker J W, 2008. An introduction to probabilistic seismic hazard analysis (PSHA). *White Paper Version 2.7.*

Bradley B A, Dhakal R P, Cubrinovski M et al., 2007. Improved seismic hazard model with application to probabilistic seismic demand analysis, *Earthquake Engineering & Structural Dynamics*, **36**(14), 2211–2225.

Bruneau M, Chang S E, Eguchi R T et al., 2003. A framework to quantitatively assess and enhance the seismic resilience of communities, *Earthquake Spectra*, **19**(4), 733–752.

Bruneau M, Reinhorn A M, 2007. Exploring the concept of seismic resilience for acute care facilities, *Earthquake Spectra*, **23**(1), 41–62.

Burton H V, Deierlein G, Lallemant D et al., 2017. Measuring the impact of enhanced building performance on the seismic resilience of a residential community, *Earthquake Spectra*, **33**(4), 1347–1367.

California Seismic Safety Commission (CSSC), 2000. *A History of the California Seismic Safety Commission: Living Where the Earth Shakes* 1975-2000. CSSC Publication 00-04, Sacramento, CA.

Chang S E, Shinozuka M, 2004. Measuring improvements in the disaster resilience of communities, *Earthquake Spectra*, **20**(3), 739–755.

Cimellaro G P, Reinhorn A M, Bruneau M, 2010a. Seismic resilience of a hospital system, *Structure & Infrastructure Engineering*, **6**(1–2), 127–144.

Cimellaro G P, Reinhorn A M, Bruneau M, 2010b. Framework for analytical quantification of disaster resilience, *Engineering Structures*, **32**(11), 3639–3649.

Cimellaro G P, Reinhorn A M, Bruneau M, 2011. Performance-based metamodel for healthcare facilities, *Earthquake Engineering & Structural Dynamics*, **40**(11), 1197–1217.

Cimellaro G P, Pique M, 2015. Seismic performance of healthcare facilities using discrete event simulation models, *Geotechnical, Geological and Earthquake Engineering*, **33**: 203–215.

Cimellaro G P, Pique M, 2016. Resilience of a hospital Emergency Department under seismic event, *Advances in Structural Engineering*, **19**(5), 825–836.

Cimellaro G P, Renschler C, Reinhorn A M et al., 2016a. PEOPLES: A framework for evaluating resilience, *Journal of Structural Engineering*, **142**(10), 04016063.

Cimellaro G P, Malavisi, M, Mahin, S, 2016b. Using discrete event simulation models to evaluate resilience of an emergency department, *Journal of Earthquake Engineering*, **21**(2), 203–226.

Cornell C A, Krawinkler H, 2000. Progress and challenges in seismic performance assessment, *PEER Center News*, **3**(2), 1–4.

Cornell C A, Jalayer F, Hamburger R O et al., 2002. Probabilistic basis for 2000 SAC Federal Emergency Management Agency Steel Moment Frame Guidelines, *Journal of Structural Engineering (ASCE)*, **128**(4), 526–533.

Department of Health and Environment (DOHE), 1995. *Actual Result of the Investigation on the Disaster Medicine.* Hanshin-Awaji Great Earthquake Restoration Headquarters, Kobe, Japan.

Dong Y, Frangopol D M, Saydam D, 2013. Time-variant sustainability assessment of seismically vulnerable bridges subjected to multiple hazards, *Earthquake Engineering & Structural Dynamics*, **42**(10), 1451–1467.

Dong Y, Frangopol D M, Saydam D, 2014. Sustainability of highway bridge networks under seismic hazard, *Journal of Earthquake Engineering*, **18**(1), 41–66.

Dong Y, Frangopol D M, 2016. Probabilistic assessment of an interdependent healthcare-bridge network system under seismic hazard, *Structure and Infrastructure Engineering*, **13**(1), 160–170.

Favier P, Poulos A, Vásquez J A et al., 2019. Seismic risk assessment of an emergency department of a Chilean hospital using a patient-oriented performance model, *Earthquake Spectra*, **35**(2), 489–512.

FEMA P695, 2009. *Quantification of Building Seismic Performance Factors*. FEMPA, Washington, DC.

FEMA P58, 2012. *Seismic Performance Assessment of Buildings*, Vols. **1** and 2: *Methodology*. Applied Technology Council (ATC), Redwood City, CA.

GB 50011-2010, 2010. *Code for Seismic Design of Buildings*. China Architecture and Building Press, Beijing (in Chinese).

GB 51039-2014, 2014. *Code for Design of General Hospital*. China Planning Press, Beijing (in Chinese).

Gunn S W A, 1995. Health effects of earthquakes, *Disaster Prevention and Management*, **4**(5), 6–10.

Hassan E M, Mahmoud H, 2018. A framework for estimating immediate interdependent functionality reduction of a steel hospital following a seismic event, *Engineering Structures*, **168**: 669–683.

Hassan E M, Mahmoud H, 2019. Full functionality and recovery assessment framework for a hospital subjected to a scenario earthquake event, *Engineering Structures*, **188**: 165–177.

HAZUS-MH MR3, 2003. *Multi-Hazard Loss Estimation Methodology-Earthquake Model*, Technical Manual. Department of Homeland Security, Washington, DC.

Hisham M, Yassin M, 1994. Nonlinear analysis of prestressed concrete structures under monotonic and cycling loads, Ph.D. Thesis. University of California, Berkeley.

Jennings P C, Housner G W, 1971. *The San Fernando, California, Earthquake of February 9, 1971*. U.S. Geological Survey and the National Oceanic and Atmospheric Administration, Reston, VA.

Jordan E, Javernick-Will A, 2013. Indicators of community recovery: Content analysis and Delphi approach, *Natural Hazards Review*, **14**(1), 21–28.

Lee S, Walsh P, 2011. SWOT and AHP hybrid model for sport marketing outsourcing using a case of intercollegiate sport, *Sport Management Review*, **14**(4), 361–369.

Liedtka S L, 2005. Analytic hierarchy process and multi-criteria performance management systems, *Journal of Cost Management*, **19**(6), 30–38.

Lin P, Wang N, Ellingwood B R, 2016. A risk de-aggregation framework that relates community resilience goals to building performance objectives, *Sustainable & Resilient Infrastructure*, **1**(**1–2**), 1–13.

Lupoi A, Cavalieri F, Franchin P, 2012. Probabilistic seismic assessment of health-care systems at regional scale. In: *Proceeding of 15th World Conference on Earthquake Engineering*, September 24-28, Lisboa, Portugal, pp. 1–10.

Malkin P V E, Semple K, 2017. Mexico earthquake, strongest in a century, Kills Dozens, *New York Times*.

Mander J B, Priestley M J N, Park R, 1988. Theoretical stress-strain model for confined concrete, *Journal of Structural Engineering* (*ASCE*), **114**(8), 1804–1826.

Miles S B, Chang S E, 2006. Modeling community recovery from earthquakes, *Earthquake Spectra*, **22**(2), 439–458.

Moehle J, Berger J, Bray J et al, 2010. *The 27 February 2010 Central South Chile Earthquake: Emerging Research Needs and Opportunities*. EERI Workshop Report, 2010, https://eeri.org/wp-content/uploads/store/Free%20PDF%20Downloads/Chile-Workshop-Report_FINAL.pdf

Myrtle R C, Masri S F, Nigbor R L et al., 2005. Classification and prioritization of essential systems in hospitals under extreme events, *Earthquake Spectra*, **21**(3), 779–802.

Office of Statewide Health Planning & Development (OSHPD), 1995. *The Northridge Earthquake*, Report 7250, Sacramento, CA.

Okoli C, Pawlowski S D, 2004. The Delphi method as a research tool: An example, design considerations and applications, *Information & Management*, **42**(1), 15–29.

Orencio P M, Fujii M, 2013. A localized disaster-resilience index to assess coastal communities based on an analytic hierarchy process (AHP), *International Journal of Disaster Risk Reduction*, **3**(1), 62–75.

Panahi M, Rezaie F, Meshkani S A, 2014. Seismic vulnerability assessment of school buildings in Tehran city based on AHP and GIS, *Natural Hazards & Earth System Science*, **1**(5), 4511–4538.

Paul J A, George S K, Yi P et al., 2006. Transient modeling in simulation of hospital operations for emergency response, *Prehospital and Disaster Medicine*, **21**(4), 223–236.

Porter K, Kennedy R, Bachman R, 2007. Creating fragility functions for performance-based earthquake engineering, *Earthquake Spectra*, **23**(2), 471–489.

Saaty T L, 1990. How to make a decision: The analytic hierarchy process, *European Journal of Operational Research*, **48**(1), 9–26.

Santarsiero G, Sarno L D, Giovinazzi S et al., 2018. Performance of the healthcare facilities during the 2016–2017 Central Italy seismic sequence, *Bulletin of Earthquake Engineering*, **4**: 1–27.

Solberg K M, Dhakal R P, Mander J B et al., 2010. Computational and rapid expected annual loss estimation methodologies for structures, *Earthquake Engineering & Structural Dynamics*, **37**(1), 81–101.

Soroushian S, Rahmanishamsi E, Ryu K P et al., 2016. Experimental fragility analysis of suspension ceiling systems, *Earthquake Spectra*, **32**(2), 150504135037002.

Teng R, Shang Q X, Zhong X et al., 2018. Fragility studies for masonry infill walls based on experimental data, *World Earthquake Engineering*, **34**(2), 96–103 (in Chinese).

Wang T, Shang Q X, Chen X et al., 2019. Experiments and fragility analyses of piping systems connected by grooved fit joints with large deformability, *Frontiers in Built Environment*, **5**(49), 1–14.

Wang X L, 2005. Seismic reliability analysis for functional utility system in hospital, Ph.D. Thesis, Beijing University of Technology, Beijing (in Chinese).

Wang Y M, Xiong L H, Xu W X, 2013. Seismic damage and damage enlightenment of medical buildings in Lushan MS 7.0 earthquake, *Earthquake Engineering and Engineering Dynamics*, **33**(4), 44–53 (in Chinese).

Welch S J, Asplin B R, Stone-Griffith S. et al., 2011. Emergency department operational metrics, measures and definitions: Results of the second performance measures and benchmarking summit, *Annals of Emergency Medicine*, **58**(1), 33–40.

Yodo N, Wang P, 2016. Engineering resilience quantification and system design implications: A literature survey, *Journal of Mechanical Design*, **138**(11), 111408.

Yu P, Wen W P, Ji D F, Zhai C H, Xie L L, 2019. A framework to assess the seismic resilience of urban hospitals, *Advances in Civil Engineering*, 7654683: 1–11.

Zhao S, Liu X, Zhuo Y, 2017. Hybrid hidden markov models for resilience metrics in a dynamic infrastructure system, *Reliability Engineering & System Safety*, **164**: 84–97.

Zhong S, Clark M, Hou X Y et al., 2015. Development of key indicators of hospital resilience: A modified Delphi study, *Journal of Health Services Research & Policy*, **20**(2), 74.

6 Seismic Resilience Assessment of Emergency Departments Based on the State Tree Method

6.1 INTRODUCTION

High mortality and casualty rates caused by major earthquakes significantly affect many aspects of society. It is estimated that 97% of earthquake-related injuries occur immediately after or within the first 30 minutes after the main shock (Gunn 1995). The local hospital response is the most important emergency response after an earthquake to provide emergency medical care. The World Health Organization (WHO 2007) stated that healthcare systems "must be physically resilient and able to remain operational and provide vital health services" after disasters. However, hospital systems are often vulnerable systems during earthquakes. Healthcare facilities were reported to have suffered great losses due to earthquakes during the last two decades (Schultz et al. 2003; UNICEF 2004; IASC 2005; PDNA 2010; Filiatrault and Sullivan 2014; Nakashima et al. 2014; Wang et al. 2013). This vulnerability renders hospitals, particularly emergency departments (EDs), less resilient to earthquake events. Before applying measures to improve seismic resilience, existing ED systems should be evaluated in terms of functionality loss, recoverability, etc.

Different frameworks have been proposed to evaluate resilience recently. Kircher et al. (2006) developed a comprehensive GIS-based technology for estimating earthquake damage and loss. The method was later extended to evaluate damage and loss of buildings subjected to hurricanes by Vickery et al. (2006). Gilbert and Ayyub (2016) developed two loss models for measuring the performance of a system from the perspective of economy. Liu et al. (2017) proposed a framework combining dynamic modeling and resilience analysis to investigate the resilience of interconnected critical infrastructures. A framework named PEOPLES for measuring community resilience at different spatial and temporal scales was proposed by Cimellaro et al. (2016b) and Kammouh et al. (2019), which combines seven different dimensions of resilience using a layered approach. Meanwhile, different resilience indices have been proposed for quantifying seismic resilience of infrastructure systems in recent publications, e.g., resilience definition by Ayyub (2015), damage index for low-rise buildings (Peng

DOI: 10.1201/9781003457459-6

et al. 2016), performance index for gas distribution network (Cimellaro et al. 2015a), resilience index for water distribution network (Cimellaro et al. 2015b), resilience and risk index at the country level (Kammouh et al. 2017), downtime model for building structures (De Iuliis et al. 2019), and infrastructure systems (Kammouh et al. 2018). For hospital system or ED, there are also some attempts to quantify resilience. Hu et al. (2012) proposed a conceptual evolutionary framework for aseismic decision support for hospitals to integrate a range of engineering and sociotechnical models. V'asquez et al. (2017) assessed the resilience of the healthcare network by comparing its normal functionality to that in the emergency and recovery phases after the 2014 Pisagua earthquake. During the resilience assessment, critical components that most significantly affect the seismic performance of the entire system can be identified (Myrtle et al. 2005; Masi et al. 2012; Miniati and Iasio 2012; Jacques et al. 2014). It was pointed out that physical damage to structural components, nonstructural components, and contents should be considered in the resilience assessment of hospitals (WHO 2006; McCabe et al. 2010; Kirsch 2010; Yavari et al. 2010; Achour et al. 2011; FEMA P58 2012). The resilience of the hospitals depends not only on the robustness of the facilities and their critical units but also on their interrelations and interactions. Although sophisticated methods for seismic resilience assessment exist, one issue that has to be carefully considered first is the definition and quantification of the functionality of a system. This remains a difficulty due to the lack of consensus on performance measures (Ellingwood et al. 2016). Performance measures including waiting time of patients (Paul et al. 2006; Cimellaro et al. 2011; Welch et al. 2011; Sørup et al. 2013; Cimellaro and Pique 2015; Cimellaro et al. 2016a; Favier et al. 2019), number of available beds (Cimellaro et al. 2011; Lupoi et al. 2012), and number of operating rooms (Cimellaro et al. 2011; Lupoi et al. 2012) are usually used to quantify functionality of EDs. These indices have real meanings for ED functionality and should be strongly supported by hospital administrations.

To quantify the system functionality, the logic relationship between the components has to be analyzed, and commonly, system analysis methods are used. Discrete event simulation (DES) is often used to predict the performance of hospitals under normal conditions (Duguay and Chetouane 2007; Gunal 2012; Best et al. 2014; Gul and Guneri 2015), and this method has been typically used to calibrate meta-models that simplify the system significantly; however, earthquake influences have not been considered. Jacques et al. (2014) assessed the functionality of critical hospital services using a fault tree method, but the fault trees were considered to be deterministic because the state of each component was based on field data. Recently, Hassan and Mahmoud (2018, 2019) extended the fault tree developed by Jacques et al. (2014) to estimate earthquake-induced functionality reduction of a hospital building. The functionality is estimated based on hospital losses that result from its sustained damage and damage to other lifelines. Recovery curves of the investigated hospital are obtained using the continuous Markov chain process. Miniati et al. (2014) developed a new integrated method based on a combination of fault tree analysis and rapid seismic vulnerability assessment. The method was then applied to two hospital systems for risk mitigation. Mieler et al. (2015) used an event tree method to translate community resilience goals into specific performance targets for important systems and components. These methods are capable of determining the logic relationship between components. However, the component influence on the system functionality is difficult to quantify through these methods.

In this study, a quantitative framework for system fragility analysis and resilience assessment of EDs based on the state tree method is proposed. The state tree model includes all success paths of the system and the component influence on the functionality of EDs is explicitly considered. System fragility is then defined, and a new metric is derived to assess the functionality and seismic resilience of EDs. The feasibility and effectiveness of the proposed method are demonstrated by conducting a functionality analysis and resilience assessment of a real ED in a hospital.

6.2 FRAMEWORK FOR SEISMIC RESILIENCE ASSESSMENT BASED ON STATE TREE METHOD

The flowchart of the proposed framework for quantifying the seismic resilience of engineering systems is illustrated in Figure 6.1. The flowchart has four blocks, including the probabilistic seismic hazard analysis (PSHA), probabilistic seismic fragility analysis (PSFA), probabilistic seismic state analysis (PSSA), and seismic resilience evaluation analysis (SREA). As indicated, the seismic hazard curve of the site is identified by the PSHA (Cornell 1968; Baker 2008). Based on the source model and attenuation model suitable for the specific seismic environment of the site of interest, the joint occurrence of the shaking intensity considering the uncertainties regarding the location, size, mechanism, and magnitude of possible future earthquakes is characterized, and several ground motions are selected for a nonlinear time history analysis (NTHA). Subsequently, a finite element model (FEM) of the system structure is created and NTHAs are conducted to obtain the engineering demand parameters (EDPs) (e.g., story drift and floor acceleration), which will be used to develop the probabilistic seismic demand models (PSDMs). PSDMs will be used in the fragility analysis of the structural components and nonstructural components to obtain the system-level component fragility curve (Tavares et al. 2012; Shang et al. 2018a). The maximum inter-story drift ratio (IDR) and peak floor acceleration (PFA) are used for the fragility analysis of different components (e.g., displacement-sensitive components and acceleration-sensitive components). A state tree model is then established for the system. The operational state of each component under a given earthquake intensity can be determined based on the results of probabilistic seismic fragility analysis and Monte Carlo simulation. The operational state of each possible

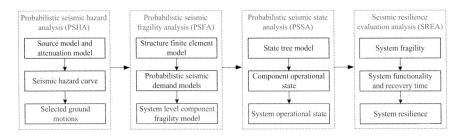

FIGURE 6.1 Overview of the quantitative framework for seismic resilience assessment.

working line is determined by assembling the state of the components using the state tree model (Li et al. 2019). System fragility is then calculated using a fitted lognormal distribution. System residual functionality under a given earthquake intensity will be calculated using the system fragility and the number of available working lines. A practical repair path that considers the component location and construction sequence is introduced to evaluate the system recovery process. Through this repair path, the functionality improvement and the recovery time will be calculated. Finally, the system recovery curve, i.e., the resilience curve, is plotted, by which the seismic resilience of the concerned system can be quantified.

6.3 SYSTEM ANALYSIS OF EMERGENCY FUNCTIONALITY

The fault tree method is one of the most commonly used methods for system performance assessment, especially in the nuclear industry (Ellingwood and Song 1996). The fault tree describes the constitution of a complex system and clearly expresses the causality between the components. Once the fault tree is established, the minimal cut set (MCS) method is used to calculate the system failure probability. The fault tree method is often used with the event tree method (Mieler et al. 2013, 2015) to describe the possible consequence. However, the failure probability of a combination of several MCSs is difficult to calculate because of the interdependency between different components. And heavy computational effort is required. Moreover, the number of system states increases significantly with the number of components. To solve these difficulties, Li et al. (2019) proposed a new probability-based method, namely the state tree method, to evaluate system performance. The proposed method includes five steps: (1) definition of system functionality; (2) system analysis using the fault tree and success path method; (3) component fragility definition; (4) evaluation of the operational states of the components and systems using the state tree model; and (5) calculation of the system failure probability using a Monte Carlo simulation or other methods such as Latin hypercube sampling. The new method explicitly considers the interdependency of all components associated with system functionality and integrates the merits of the fault tree method and the success path method (Electric Power Research Institute 1991). Unlike the seismic marginal assessment method (Electric Power Research Institute 1991), all success paths have to be identified and clearly defined to construct the system model in the state tree method. A strong logical relationship between the components inherently exists within the state tree model, and it is much more manageable than the minimal cut set method used in fault tree analysis.

6.3.1 SYSTEM MODELING OF EMERGENCY FUNCTIONALITY

The emergency functionality of the ED in a hospital is comprised of three sub-functionalities, i.e., the basic functionality (BF), medical examination functionality (MEF), and medical treatment functionality (MTF). All three sub-functionalities are essential to preserve emergency functionality. The MEF ensures that patients can be checked by doctors as soon as they arrive at the hospital after the earthquake; the MTF helps provide medical treatment for patients in time and the BF ensures that the medical care is available and the environment is acceptable.

The failure of either of these three sub-functionalities would result in the failure of the emergency functionality. The fault trees of the three sub-functionalities are illustrated in Figure 6.2. BF is supported by several subsystems including the structural system, the transportation system, the enclosure system, the power supply system, and the water supply system (Figure 6.2). The structural system ensures the safety of the ED and people inside the building. The transportation system ensures that injured people have access to doctors to get medical treatment. The enclosure system helps create an aseptic environment for patients. The water supply system and power supply system provide essential working conditions for the ED. Both water and power supplies inside and outside the hospital are considered. The MEF is divided into the X-ray room and clinical laboratory for the diagnosis. It is assumed that the examination work can be conducted if either the X-ray room or clinical laboratory is accessible; an "AND" gate is used here (Figure 6.2). The MTF is divided into the transfusion room treatment and operating room treatment. It is assumed that the treatment can be conducted in either the transfusion room or the operating room; an "AND" gate is used to describe the relationship (Figure 6.2). The bottom events include different components that are used to support diagnosis and treatment.

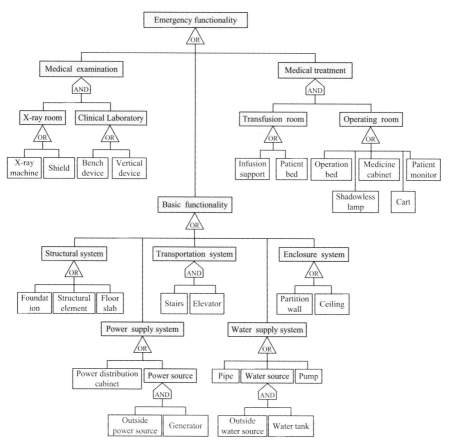

FIGURE 6.2   Fault tree of the emergency functionality.

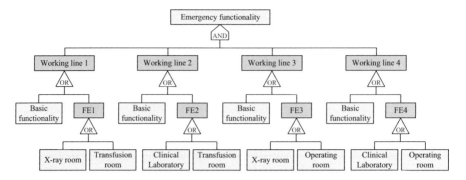

FIGURE 6.3   State tree of emergency functionality.

For the emergency functionality, the state tree model is developed based on the fault tree model as mentioned above. Several fictional events (FEs) are added to represent the states of the component groups (Figure 6.3). FE1, FE2, FE3, and FE4 represent four pairwise combinations of MEF and MTF; each can provide emergency functionality accompanied by BF, as shown in Figure 6.3. The state tree model will be used in probabilistic seismic state analysis as depicted in Figure 6.1. Each pairwise combination of the BF with the FE1, FE2, FE3, and FE4 is called a success path (also called working line in this study) and provides medical services to patients or injured people. It is worth noting that the staff behavior (e.g., doctors and nurses) in the ED is not included in the state tree model.

The BF, X-ray room, transfusion room, clinical laboratory, and operating room are referred to as functional units in this study and the bottom components of these functional units are independent. Therefore, the failure and success probability of the five functional units can be calculated based on the aforementioned rules. The five functional units are also independent of each other. This calculation method is based on the failure probability of each independent component and is referred to as the direct method in this study. The results of the direct method are exact considering that no approximations are made in this method.

6.3.2 DEFINITION OF SYSTEM FRAGILITY AND FUNCTIONALITY

The functionality index F represents the availability of the working lines as depicted in Figure 6.3. To be consistent with the definition of component fragility, the system fragility $P(F|IM)$ is defined as the exceedance probability of a maximum of N working lines surviving at a given earthquake intensity measure (IM) (the peak ground acceleration (PGA) is used as the earthquake intensity measure in this study). The values for F are 0%–75% in increments of 25%, where $F=0\%$ means that none of the lines work, $F=25\%$ means that a maximum of one line works, $F=50\%$ means that no more than two lines succeed, and $F=75\%$ means that a maximum of three lines work. $P[N = n \,|\, \text{PGA} = \text{pga}]$ is the probability that the number of working lines available is i at a given PGA; it is the difference between the probabilities of exceeding a maximum of i and $i+1$ working lines in Equation (6.1). Based on

Equation (6.1), $P(F|IM)$ can also be calculated if $P[N = n \mid PGA = \text{pga}]$ is known for the system:

$$P[N = n \mid PGA = \text{pga}] = \begin{cases} 1 - P(F = 75\% \mid PGA = \text{pga}) & n = 4 \\ P(F = 75\% \mid PGA = \text{pga}) - P(F = 50\% \mid PGA = \text{pga}) & n = 3 \\ P(F = 50\% \mid PGA = \text{pga}) - P(F = 25\% \mid PGA = \text{pga}) & n = 2 \\ P(F = 25\% \mid PGA = \text{pga}) - P(F = 0\% \mid PGA = \text{pga}) & n = 1 \\ P(F = 0\% \mid PGA = \text{pga}) & n = 0 \end{cases}$$

(6.1)

Hospital systems are designed to provide medical care for patients. The emergency functionality of the ED is defined as the ability to provide emergency medical care and is calculated as the normalized number of working lines (percentage of full ED operation) as defined in Equation (6.2), where N_{total} is the number of possible working lines. This represents the reference functionality before the earthquake or under normal operational conditions, and N_{total} is four in this study; $N(t)$ is the number of working lines at a given PGA and is calculated based on the full probability theory as defined in Equation (6.3); $Q(t)$ represents the functionality of the system at time t. Equation (6.2) will be used to quantify system functionality and be involved in resilience evaluation as depicted in Figure 6.1.

$$Q(t) = \frac{N(t)}{N_{total}}$$

(6.2)

$$N(t) = \sum_{i=0}^{N_{total}} i \cdot P[N = i \mid PGA = \text{pga}]$$

(6.3)

6.3.3 MONTE CARLO SIMULATION

It is very difficult to use the direct method (Section 6.3.1) when the components or FEs are coupled with each other in the state tree method or fault tree method for large and complex systems. Therefore, a Monte Carlo simulation is used to calculate the system fragility in this study to broaden the use of the state tree method. In one simulation, a random number uniformly distributed between 0 and 1 is generated for each bottom component of the state tree model. If the generated number is less than or equal to the failure probability ($P(C)$) defined by the corresponding component fragility curve at a given PGA, the component is considered to have failed and the recovery time is determined by the component recovery model; otherwise, it succeeds and the recovery time for this component is 0 days. For example, if the failure probability of a component is 0.45 at a PGA of 0.5 g and the corresponding random number is 0.17, then the component will fail. In this study, for simplicity, the recovery model of each component is determined by the assumed median recovery time needed to repair the component.

As the framework shown in Figure 6.1, once the states of all components are determined during one simulation, the system operational state can be calculated using the state tree model, and the recovery time of each repair sequence (Section 6.4.3), the system recovery time, and the functionality loss can be determined.

For a given PGA, the simulation is repeated N_{sim} times using different random numbers, and the system fragility defined in Section 6.3.2 is then calculated as the ratio of the numbers for different F ($N \leq 0$, 1, 2, 3) to the number of simulations (N_{sim}) using Equations (6.4) and (6.1), where $N_{success}$ is the number of success samples of the working lines at a specific number (0, 1, 2, 3, 4). The ratio of the sum of the recovery time and functionality improvement of each repair sequence during the N_{sim} times simulations to the number of simulations (N_{sim}) is used to obtain the mean resilience curve (Cimellaro et al. 2010a, b). The mean recovery time of the system is determined by summing up the values of each repair sequence.

$$P(N = n \mid PGA = \mathrm{pga}) = \frac{1}{N_{sim}} \left[N_{success} \middle| (N = n, PGA = \mathrm{pga}) \right] \qquad n = 0,1,2,3,4$$

(6.4)

As explained above, the simulation process is repeated m times for the ground motions of different intensities in order to construct the fragility curve of the system. Discrete points of the fragility model of the system are plotted and nonlinearly fitted using MATLAB (The MathWorks, Inc. 2016) with the Hougen-Watson model (Bates and Watts 1988). It is worth noting that in every simulation, the states of the components, FEs, and the system are uniquely determined, and the influences of the shared components are considered in the proposed method.

6.4   CASE STUDY OF EMERGENCY FUNCTIONALITY IN A HOSPITAL

6.4.1   Introduction to the Hospital Building

The hospital building used for the case study was a 10-story reinforced concrete (RC) frame designed following the Chinese Code for Seismic Design of Buildings (GB 50011-2010, 2010). The concrete was C30 with a standard compressive strength of 20.1 MPa, and the steel rebars were HRB400 with a yielding strength of 360 MPa. The three-dimensional model and plan view of the first floor are shown in Figure 6.4. A finite element model of the hospital building was created by OpenSees. The beams and columns were simulated using nonlinear beam-column elements, which were defined as force-based elements with distributed plasticity. The composite RC cross-sections were conveniently simulated by the fiber formulation. In the finite element model, the story mass and the corresponding gravity were uniformly distributed to each beam-column joint, and the rigid diaphragm constraint was used for each floor. A geometric nonlinearity was considered for the P-Delta effect. The first three fundamental natural periods of the building were 1.2362, 1.1599, and 1.0772 s, respectively.

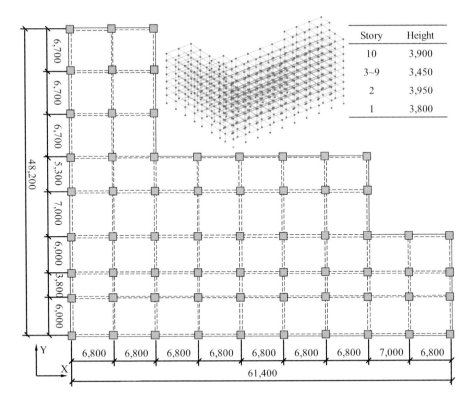

Story	Height
10	3,900
3~9	3,450
2	3,950
1	3,800

FIGURE 6.4   Plan view and three-dimensional model of the hospital building used for the case study (unit: mm).

Without the background information regarding earthquakes and geologic information on the area where the building is located, the PSHA cannot be conducted. Instead, the far-field record set that includes 22 records was selected from the Pacific Earthquake Engineering Research Center Next Generation Attenuation (PEER NGA) database (Ancheta et al. 2013), and we used the criteria suggested by FEMA P695 (FEMA 2009), which were adopted to evaluate the seismic performance of the hospital building. These 22 records were obtained from 14 events that occurred between 1971 and 1999. More information about the ground motions can be found in Appendix A.9 of FEMA P695 (FEMA 2009).

The engineering demand parameter (EDP) (IDR, PFA, etc.) used to define the fragility model (Porter et al. 2007) of a specific component in a specific structure is transformed into an earthquake intensity measure [such as the PGA or S_a (T_1, 5%)]. The transformation method is based on the probabilistic seismic demand analysis (PSDA) (Cornell et al. 2002; Tavares et al. 2012). Parameters used to define the IDR and PFA of each story were obtained from PSDA and shown in Table 6.1. With the PSDM of the component (more detailed information is available in Shang et al. 2018b) and the capacity as defined by the component fragility curve (Porter et al. 2007), the system-level component fragility is defined as the probability that a component in a typical

TABLE 6.1

Parameters of the Probabilistic Seismic Demand Model

Number	Maximum Inter-Story Drift Ratio				Peak Floor Acceleration			
of Story	b	$\ln(a)$	R^2	$\beta_{D|IM}$	b	$\ln(a)$	R^2	$\beta_{D|IM}$
1	0.9281	–4.3225	0.7849	0.2732	0.9383	–0.6160	0.7338	0.2360
2	0.8324	–3.9517	0.7983	0.2353	0.8893	–0.0366	0.8070	0.1816
3	0.8770	–3.7688	0.8341	0.2200	0.9403	0.2604	0.8753	0.1482
4	0.9129	–3.6973	0.8617	0.2057	1.0084	0.4192	0.9037	0.1374
5	0.8988	–3.7879	0.8644	0.2002	0.9987	0.4145	0.9023	0.1372
6	0.8467	–4.0145	0.8209	0.2224	0.9399	0.3161	0.8591	0.1589
7	0.8241	–4.2328	0.7679	0.2548	0.8851	0.2952	0.8256	0.1699
8	0.9670	–3.7654	0.5318	0.5103	0.7601	0.1988	0.7277	0.1941
9	0.6570	–4.5230	0.4051	0.4478	0.8097	0.4373	0.7991	0.1695
10	0.5797	–4.9178	0.3252	0.4696	0.9143	0.8476	0.8086	0.1857

Note: a and b are regression coefficients, R^2 is the coefficient of determination indicating the robustness of the regression model, and $\beta_{D|IM}$ is the conditional logarithmic standard deviation of PSDM (Tavares et al. 2012).

building reaches or exceeds a damaged state at a given ground motion intensity. The fragility model that used PGA as an intensity measure for the considered component is then obtained. The IDR and PFA from the NTHA are used as inputs to generate PSDMs for different components. The floor distribution, fragility definition, median recovery time, and system-level component fragility definition of the components used in the system analysis of the emergency functionality are given in Table 6.2.

6.4.2 System Fragility Analysis

The system fragility analysis was conducted using the direct method proposed in Section 6.3.1 and the Monte Carlo simulation method proposed in Section 6.3.3. For the direct method, the total number of combinations of the four working lines is two (number of states) raised to the power equal to the number of working lines (four in this study), $2^4 = 16$. Table 6.3 shows the 16 combinations. According to the state tree (Figure 6.3), the BF needs to succeed to ensure the success of the emergency functionality. The failure and success probability of the BF are calculated using the rules of fault tree. Let P_1, P_2, P_3, P_4, and P_5 represent the failure probabilities of the BF, X-ray room, clinical laboratory, transfusion room, and operating room, respectively; the probability of occurrence of the combinations is calculated as depicted in Table 6.3. The number of working lines and the specific lines that are working are obtained based on the probabilities, as depicted in Figure 6.5. When there are two working lines, the probability of working line 1 and working line 4 succeeding simultaneously is zero and the same result is obtained for the case of working line 2 and working line 3 succeeding. There are four working lines if working line 1 and working line 4 succeed simultaneously, which means that working line 2 and working line 3 also succeed at the same time. The probability of occurrence curve of working line 1 and working line 3 succeeding coincides

TABLE 6.2

Component Distribution and Fragility Definition (units: g, day)

Number	Component	Floor Distribution	EDP	Median	Dispersion	Recovery Time	Reference	Median (PGA)	Dispersion (PGA)
1	Foundation	–	–	–	–	–	–	–	–
2	Structural element	1–10	IDR	0.04	0.4	37.08	HAZUS-MH MR3 (2003)	0.97	0.5563
3	Floor slab	1–10	IDR	0.04	0.25	16.2	FEMA P58 (2012)	0.97	0.4449
4	Stairs	1–10	IDR	0.028	0.45	16.2	FEMA P58 (2012)	0.66	0.5984
5	Elevator	1–10	PFA	0.5	0.3	12.96	FEMA P58 (2012)	0.19	0.3859
6	Partition wall	1–10	IDR	0.0171	0.4	2	Pali et al. (2018)	0.46	0.5563
7	Ceiling	1–10	PFA	1.791	0.419	7.56	Soroushian et al. (2016)	0.75	0.5012
8	Pipe	1–10	PFA	2.25	0.5	9	FEMA P58 (2012)	0.96	0.5833
9	Pump	–	PGA	2.25	0.5	6.84	FEMA P58 (2012)	2.25	0.5
10	Outside water source	–	PGA	2.25	0.5	5	FEMA P58 (2012)	2.25	0.5
11	Water tank	10	PFA	2.16	0.45	5	FEMA P58 (2012)	0.65	0.5324
12	Power distribution cabinet	–	PGA	2.16	0.45	8.28	FEMA P58 (2012)	2.16	0.45
13	Generator	–	PGA	2.25	0.5	5	FEMA P58 (2012)	2.25	0.5
14	Power source	–	PGA	2.25	0.5	2	FEMA P58 (2012)	2.25	0.5
15	X-ray machine	2	PFA	2.16	0.45	5.4	FEMA P58 (2012)	1.76	0.5456
16	Shield	2	IDR	0.02	0.4	11.88	Wang (2005)	0.75	0.5575
17	Bench device	2	PFA	2.16	0.45	2	FEMA P58 (2012)	1.76	0.5456
18	Vertical device	2	PFA	2.16	0.45	2	FEMA P58 (2012)	1.76	0.5456
19	Infusion support	2	PFA	2.16	0.45	2	FEMA P58 (2012)	1.76	0.5456
20	Patient bed	2	PFA	2.16	0.45	2	FEMA P58 (2012)	1.76	0.5456
21	Operation bed	6	PFA	2.16	0.45	2	FEMA P58 (2012)	1.15	0.5078
22	Patient monitor	6	PFA	2.16	0.45	2	FEMA P58 (2012)	1.15	0.5078
23	Shadowless lamp	6	PFA	2.16	0.45	2	FEMA P58 (2012)	1.15	0.5078
24	Cart	6	PFA	2.16	0.45	2	FEMA P58 (2012)	1.15	0.5078
25	Medicine cabinet	6	PFA	2.16	0.45	2	FEMA P58 (2012)	1.15	0.5078

EDP, engineering demand parameter; IDR, inter-story drift ratio; PFA, peak floor acceleration; PGA, peak ground acceleration.

TABLE 6.3
Operational States of the Functional Units and Working Lines ($\sqrt{}$ Means Success and $\times$ Means Failure)

		Operational State of Functional Units						
	Basic Functionality	X-ray Room	Clinical Laboratory	Transfusion Room	Operating Room	Number of Working Lines	Working Line	Probability of Occurrence
1	$\sqrt{}$	$\times$	$\times$	$\times$	$\times$	0	–	$P=(1-P_1)\cdot P_2\cdot P_3\cdot P_4\cdot P_5$
2	$\sqrt{}$	$\times$	$\times$	$\times$	$\sqrt{}$	0	–	$P=(1-P_1)\cdot P_2\cdot P_3\cdot P_4\cdot(1-P_5)$
3	$\sqrt{}$	$\times$	$\times$	$\sqrt{}$	$\times$	0	–	$P=(1-P_1)\cdot P_2\cdot P_3\cdot(1-P_4)\cdot P_5$
4	$\sqrt{}$	$\times$	$\sqrt{}$	$\times$	$\times$	0	–	$P=(1-P_1)\cdot P_2\cdot(1-P_3)\cdot P_4\cdot P_5$
5	$\sqrt{}$	$\sqrt{}$	$\times$	$\times$	$\times$	0	–	$P=(1-P_1)\cdot(1-P_2)\cdot P_3\cdot P_4\cdot P_5$
6	$\sqrt{}$	$\sqrt{}$	$\sqrt{}$	$\times$	$\times$	0	–	$P=(1-P_1)\cdot(1-P_2)\cdot(1-P_3)\cdot P_4\cdot P_5$
7	$\sqrt{}$	$\sqrt{}$	$\times$	$\sqrt{}$	$\times$	1	1	$P=(1-P_1)\cdot(1-P_2)\cdot P_3\cdot(1-P_4)\cdot P_5$
8	$\sqrt{}$	$\sqrt{}$	$\times$	$\times$	$\sqrt{}$	1	3	$P=(1-P_1)\cdot(1-P_2)\cdot P_3\cdot P_4\cdot(1-P_5)$
9	$\sqrt{}$	$\times$	$\sqrt{}$	$\sqrt{}$	$\times$	1	2	$P=(1-P_1)\cdot P_2\cdot(1-P_3)\cdot(1-P_4)\cdot P_5$
10	$\sqrt{}$	$\times$	$\sqrt{}$	$\times$	$\sqrt{}$	1	4	$P=(1-P_1)\cdot P_2\cdot(1-P_3)\cdot P_4\cdot(1-P_5)$
11	$\sqrt{}$	$\times$	$\times$	$\sqrt{}$	$\sqrt{}$	0	–	$P=(1-P_1)\cdot P_2\cdot P_3\cdot(1-P_4)\cdot(1-P_5)$
12	$\sqrt{}$	$\times$	$\sqrt{}$	$\sqrt{}$	$\sqrt{}$	2	2, 4	$P=(1-P_1)\cdot P_2\cdot(1-P_3)\cdot(1-P_4)\cdot(1-P_5)$
13	$\sqrt{}$	$\sqrt{}$	$\times$	$\sqrt{}$	$\sqrt{}$	2	1, 3	$P=(1-P_1)\cdot(1-P_2)\cdot P_3\cdot(1-P_4)\cdot(1-P_5)$
14	$\sqrt{}$	$\sqrt{}$	$\sqrt{}$	$\times$	$\sqrt{}$	2	3, 4	$P=(1-P_1)\cdot(1-P_2)\cdot(1-P_3)\cdot P_4\cdot(1-P_5)$
15	$\sqrt{}$	$\sqrt{}$	$\sqrt{}$	$\sqrt{}$	$\times$	2	1, 2	$P=(1-P_1)\cdot(1-P_2)\cdot(1-P_3)\cdot(1-P_4)\cdot P_5$
16	$\sqrt{}$	$\sqrt{}$	$\sqrt{}$	$\sqrt{}$	$\sqrt{}$	4	1, 2, 3, 4	$P=(1-P_1)\cdot(1-P_2)\cdot(1-P_3)\cdot(1-P_4)\cdot(1-P_5)$

exactly with that of working line 3 and working line 4 succeeding, as depicted in Figure 6.5(b). This is only a coincidence because the fragility parameters of the related components, i.e., that of the bench device, vertical device, infusion support, and patient bed are the same in this study. The same fragility parameters provide the same calculation results for $P_3 \cdot (1 - P_4)$ and $(1 - P_3) \cdot P_4$.

In the Monte Carlo simulation, the PGA value ranged from 0.01 to 1.0 g with an interval of 0.01 g, and the simulation was repeated 1,000 times (100 times were also conducted to compare the results) for each PGA value. The fragility curves (obtained from 1,000 simulations) for $F=0\%$, 25%, 50%, and 75% are shown in Figure 6.6. The fragility parameters fitted using MATLAB (The MathWorks, Inc. 2016) are also presented in Figure 6.6. The median capacities with a 50% failure probability for $F=0\%$, 25%, 50%, and 75% are all less than the median capacity of the single components, which indicates that the seismic reliability of the emergency functionality is lower than that of the individual components.

The probability of occurrence for different numbers of working lines is calculated using the direct method, as shown in Table 6.3. The probability can also be calculated

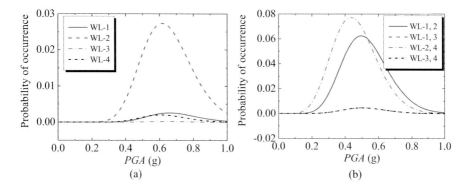

FIGURE 6.5 Probability of occurrence for the working lines. (a) $N=1$ and (b) $N=2$ (the curves of WL-1, 3 and WL-3, 4 coincide exactly in b).

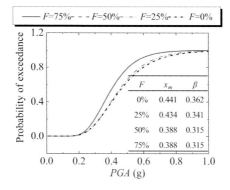

FIGURE 6.6 System fragility curves (F is the functionality index as defined in Section 3.2, x_m denotes the median value of the distribution, and β denotes the logarithmic standard deviation. The curves of $F=75\%$ and $F=50\%$ coincide exactly).

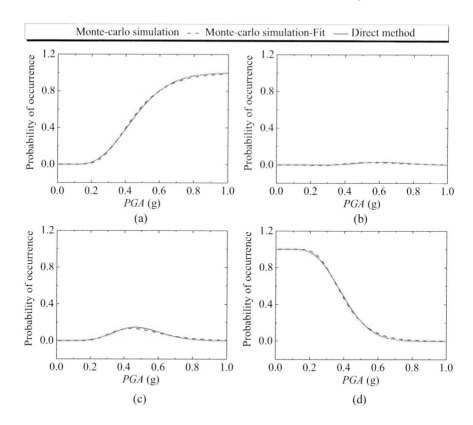

FIGURE 6.7 Probability of occurrence for different number of working lines (100 simulations) for (a) $N=0$, (b) $N=1$, (c) $N=2$, and (d) $N=4$.

by using Equation (6.1) and the Monte Carlo simulation. The results of the two methods are presented in Figure 6.7 (100 simulations) and Figure 6.8 (1,000 simulations). The results are almost the same, which indicates the feasibility and accuracy of the state tree method based on the Monte Carlo simulation. It is also demonstrated that the number of simulations has a large influence on the results. To be more specific, a smaller number of simulations provide results with more randomness and uncertainty. However, an analysis of the suitable number of simulation times is beyond the scope of this study and the number is likely different for different systems. The results of the 1,000 Monte Carlo simulations are used for the emergency functionality calculation and recovery time assessment in Sections 6.4.4 and 6.4.5.

6.4.3 PRACTICAL REPAIR PATH

Seismic resilience is a dynamic process of functionality variation over time for an engineering system after earthquakes. Functionality reduction will occur as a result of component damage during the earthquake. Subsequently, the functionality will increase when the damaged components of the system have been repaired. The

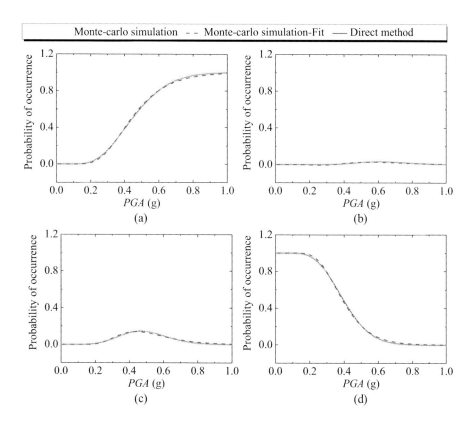

Monte-carlo simulation   - - Monte-carlo simulation-Fit   —— Direct method

FIGURE 6.8 Probability of occurrence for different number of working lines (1,000 simulations) for (a) $N=0$, (b) $N=1$, (c) $N=2$, and (d) $N=4$.

functionality reaches 1.0 (system performance level prior to the earthquake) when all damaged components have been repaired. The repair path of a system can be complex considering the construction management strategy. For simplicity, in this study, a practical repair path that considers the component location and construction sequence is used. As depicted in Figure 6.9, the practical repair path has seven steps that occur in series; in each step, several component repairs proceed in parallel. The recovery time of one parallel repair sequence is determined by the component that requires the largest time to restore. Once the recovery time of each step is obtained, the recovery time of the system is calculated as the sum of the recovery times of all seven steps. It is assumed that the recovery work is conducted in all stories simultaneously, and the recovery time of one specific component is the maximum time needed for repairing that component in each story.

6.4.4 Functionality Loss and Recovery Time Analysis

The residual functionality $Q(t=t_0)$ is the functionality after an earthquake but prior to the recovery process. The residual functionality was calculated using Equation (6.2)

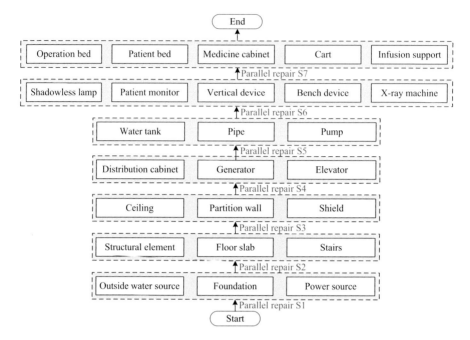

FIGURE 6.9   Practical repair path for emergency functionality.

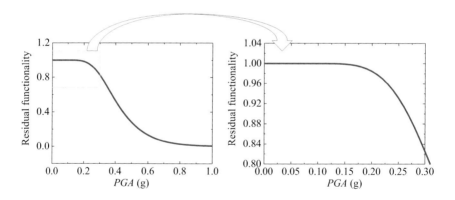

FIGURE 6.10   Residual functionality for different PGA values.

and is shown in Figure 6.10. The residual functionality starts to drop at a relatively low earthquake intensity (i.e., PGA = 0.09 g) as a result of damage to some vulnerable components (see Table 6.2). The functionality of the system after each recovery process is presented in Figure 6.11a, where Fi represents the functionality when repair sequences $S1, S2, \ldots$ and S_i ($i = 1, 2, \ldots, 6, 7$) (as defined in Section 6.4.3) are completed and F0 means no recovery, i.e., the residual functionality after an earthquake. The functionality improvement and recovery time of each repair sequence are depicted in Figure 6.11b and c. It is worth noting that it is just a coincidence that the total recovery time versus PGA

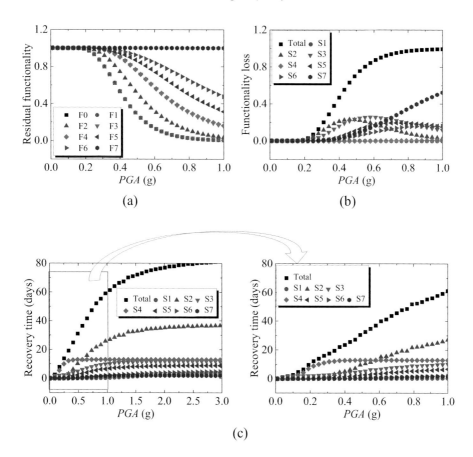

FIGURE 6.11 Functionality and recovery time after each recovery process: (a) residual functionality of the recovery process, (b) functionality improvement of each repair sequence, and (c) recovery time of each repair sequence.

is nearly linear in the range of 0.0–1.0 g in Figure 6.11c. The recovery time of the system will approach the reconstruction time of the system if all components are damaged.

The functionality loss and recovery time of each repair sequence for a maximum considered earthquake (MCE) in the Chinese code (GB50011-2010 2010) is shown in Figure 6.12. Sequences 2 and 3 contributed most to the functionality loss compared with other repair sequences. Sequence 2 includes the recovery of structural element, floor slab, and stairs, while Sequence 3 includes the recovery of ceiling, partition wall, and shield. Sequence 4 includes the recovery of distribution cabinet, generator, and elevator and had the longest recovery time because the elevator in S4 is easy to be damaged (0.19 g, see Table 6.2) and its recovery requires a long time (12.96 days, see Table 6.2).

The functionality influence of different components is identified based on the assumption that a specific component is never damaged. For example, component 4 is assumed to be robust enough to resist earthquakes and will never be damaged.

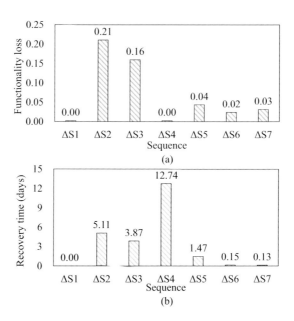

FIGURE 6.12 Functionality loss and recovery time of each repair sequence at a maximum considered earthquake (MCE): (a) functionality loss and (b) recovery time.

Based on this assumption, functionality analysis will be conducted for the system. This process will be repeated for all 25 components and different earthquake intensities. Then, the one with the biggest change from the original value of functionality is identified as the most influential component. The results are depicted in Figure 6.13. It is observed that the stairs (Component 4) and elevator (Component 5) have the largest influence on the functionality of the ED. The structural element (Component 2), partition wall (Component 6), ceiling (Component 7), pipes (Component 8), and shield (Component 16) also have large influences on the functionality. These results are consistent with the questionnaire survey results obtained by Shang et al. (2018b), who found that the structural element, stairs, elevator, infilled wall, and pipeline were some of the most important components in hospital systems. The functionality of the ED could be improved significantly if the seismic performance of these key components was improved before the occurrence of earthquakes by using new retrofitted technologies.

The residual functionality after the recovery of the sub-functionalities (Section 6.3.1) is shown in Figure 6.14a, where F-0 means no recovery, F-1 means that BF has been recovered, F-2 means that BF and MEF have been recovered, and F-3 means that BF, MEF, and MTF have been recovered. Based on the data in Figure 6.14a, the sub-functionality loss at different earthquake intensities is calculated and is presented in Figure 6.14b. The sub-functionality loss after an MCE is presented in Figure 6.14c. The BF contributes most to the functionality loss. The functionality losses of MEF and MTF are quite small after an MCE; this is attributed to the fragility model of the components that were used to support the MEF and MTF (Table 6.2).

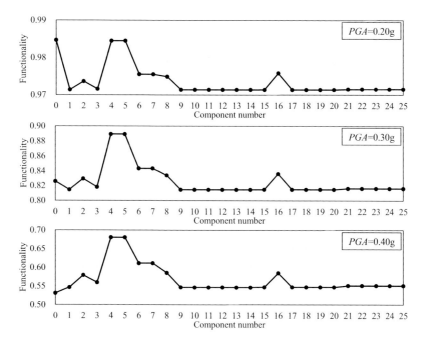

FIGURE 6.13 Influence of different components on system functionality.

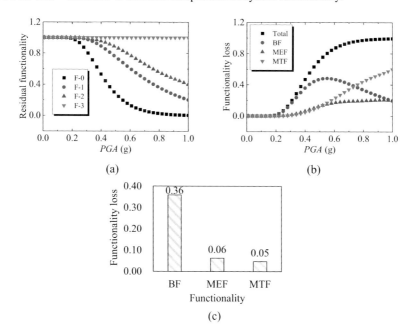

FIGURE 6.14 Sub-functionality influence on functionality loss: (a) residual functionality after recovery of each sub-functionality, (b) functionality loss resulting from each sub-functionality, and (c) sub-functionality loss after maximum considered earthquake (MCE).

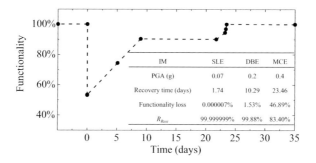

FIGURE 6.15 Resilience curve for a maximum considered earthquake (MCE).

The most critical components that easily suffer damage in earthquakes or have a large influence on the functionality of the ED can thus be identified. This allows hospital administrators, emergency planners, and staff to perform retrofitting before earthquakes and develop medical treatment plans or emergency plans that can be implemented after earthquakes.

6.4.5 RESILIENCE CURVE

Using the linear recovery model suggested by Cimellaro et al. (2010a), the resilience curves for different earthquake intensities were developed based on the repair paths, and the result for a maximum considered earthquake is presented in Figure 6.15. The resilience index R_{Resi} is determined using Equation (6.5) as suggested by Cimellaro et al. (2010b), where $Q(t)$ is the system functionality, t_{OE} is the time of occurrence of the earthquake, and T_{RE} is the recovery time. The resilience indices of different earthquake intensities are shown in Figure 6.15. After an earthquake, the functionality of the system is affected and decreases by 0.000007%, 1.53%, and 46.89% for a service level earthquake (SLE), design-basis earthquake (DBE), and MCE, respectively. The recovery time of the emergency functionality meets the 8.13-day resilience demand (Shang et al. 2018b) for the SLE intensity. However, the resilience of the emergency functionality still needs to be improved to meet the demand for DBE and MCE intensities.

$$R_{Resi} = \frac{1}{T_{RE}} \int_{t_{OE}}^{t_{OE}+T_{RE}} Q(t)dt \tag{6.5}$$

6.5 SUMMARY AND CONCLUSIONS

In this study, we proposed a framework for quantifying the seismic resilience of engineering systems. A new method called the state tree method that explicitly considers the component contribution to the functionality of the ED was integrated into the framework. The state tree model was developed based on the fault tree model and success path method. The emergency functionality and system fragility of the ED were defined and quantified. A practical repair path that took into account the component location and repair sequence was used to calculate the system recovery time.

The proposed methodological framework was used in a case study of the emergency functionality analysis and the resilience assessment of an ED in a hospital. The system fragility analyses were conducted using both the direct method (theoretical method) and the Monte Carlo simulation method. The results of both methods were in good agreement, which validated the proposed framework and the state tree model.

The recoverability of the system was calculated by using a practical repair model. The system residual functionality after the repair of the components was evaluated for the repair sequence to obtain the functionality loss and the recovery time. The influence of each component on the entire system functionality was quantified in order to identify the most important components. The seismic resilience of the ED after an MCE was quantified using a resilience curve; the results demonstrated that the resilience of the ED required improvements.

In conclusion, the proposed methodology can be used to conduct fragility analyses and resilience assessments of the emergency functionality of hospital systems. The results allow hospital administrators, emergency planners, and staff to make informed decisions regarding disaster prevention and reduction. The proposed framework can be easily applied to other engineering systems as well. Since this is a tentative study, this work bears some limitations and there are still several extensions to be explored as follows:

1. The staff behavior (e.g., doctors and nurses) in the ED and ambulance availability after earthquakes are not included in the state tree model. Moreover, outside factors including the operational state of the transportation system (e.g., roads and bridges), labor availability, and labor skill are not considered in this study. These factors are believed to have a significant influence on the functionality of ED and shall be considered by the upgraded model in the future study.
2. Each component has only two states, i.e., success or failure, in the fragility analysis. The relationship between different damage states (i.e., slight damage, moderate damage, extensive damage, and complete damage) with the functionality loss still needs to be calibrated.
3. Although some assumptions were made in the proposed framework, it is believed that the precise modeling of the ED and accurate evaluation of the emergency functionality are possible if more practical data are used.

REFERENCES

Achour N, Miyajima M, Kitaura M, Price A. 2011. Earthquake-induced structural and non-structural damage in hospitals, *Earthquake Spectra,* **27**(3), 617–634.

Ancheta T, Darragh R, Stewart J et al. 2013. *PEER NGA-West2 Database*, Technical Report PEER 2013/03. Pacific Earthquake Engineering Research Center, Berkeley, CA.

Ayyub B M. 2015. Practical resilience metrics for planning design decision making, *ASCE-ASME Journal of Risk and Uncertainty in Engineering Systems Part A: Civil Engineering,* **1**(3), 4015008.

Baker J W. 2008. An introduction to probabilistic seismic hazard analysis (PSHA). *White Paper* 2008.

Bates D M, Watts D G. 1988. *Nonlinear Regression Analysis and its Applications*. John Wiley & Sons Inc, Hoboken, NJ.

Best A M, Dixon C A, Kelton W D, Lindsell C J, Ward M J. 2014. Using discrete event computer simulation to improve patient flow in a Ghanaian acute care hospital, *The American Journal of Emergency Medicine,* **32**(8), 917–922.

Cimellaro G P, Reinhorn A M, Bruneau M. 2010a. Seismic resilience of a hospital system, *Structure & Infrastructure Engineering,* **6**(1–2), 127–144.

Cimellaro G P, Reinhorn A M, Bruneau M. 2010b. Framework for analytical quantification of disaster resilience, *Engineering Structures,* **32**(11), 3639–3649.

Cimellaro G P, Reinhorn A M, Bruneau M. 2011. Performance-based metamodel for healthcare facilities, *Earthquake Engineering & Structural Dynamics,* **40**(11), 1197–1217.

Cimellaro G P, Pique M. 2015. Seismic performance of healthcare facilities using discrete event simulation models, *Geotechnical Geological and Earthquake Engineering,* **33**, 203–215.

Cimellaro G P, Villa O, Bruneau M. 2015a. Resilience-based design of natural gas distribution networks, *Journal of Infrastructure Systems,* **21**(1), 05014005.

Cimellaro G P, Tinebra A, Renschler C et al. 2015b. New resilience index for urban water distribution networks, *Journal of Structural Engineering,* **142**(8), C4015014.

Cimellaro G P, Malavisi M, Mahin S. 2016a. Using discrete event simulation models to evaluate resilience of an emergency department, *Journal of Earthquake Engineering,* **21**(2), 203–226.

Cimellaro G P, Renschler C, Reinhorn A M et al. 2016b. PEOPLES: A framework for evaluating resilience, *Journal of Structural Engineering,* **142**(10), 04016063.

Cimellaro G P, Pique M. 2016. Resilience of a hospital emergency department under seismic event, *Advances in Structural Engineering,* **19**(5), 825–836.

Cornell C A. 1968. Engineering seismic risk analysis, *Bulletin of the Seismological Society of America,* **58**(5), 1583–1606.

Cornell C A, Jalayer F, Hamburger R O, Douglas A F. 2002. Probabilistic basis for 2000 SAC Federal Emergency Management Agency steel moment frame guidelines, *Journal of Structural Engineering,* **128**(4), 526–533.

De Iuliis M, Kammouh O, Cimellaro G P. et al, 2019. Downtime estimation of building structures using fuzzy logic, *International Journal of Disaster Risk Reduction,* **34**, 196–208.

Duguay C Chetouane F. 2007. Modeling and improving emergency department systems using discrete event simulation, *Simulation* **83**(4), 311–320.

Ellingwood B, Song J. 1996. *Impact of Structural Aging on Seismic Risk Assessment of Reinforced Concrete Structures in Nuclear Power Plants*, NUREG/CR-6425, ORNL/TM-13 149. Nuclear Regulatory Commission, Rockville, MD.

Ellingwood B, van de Lindt J W, McAllister T P. 2016. Developing measurement science for community resilience assessment: Preface to the special issue of sustainable and resilient infrastructure on the Centerville Testbed, *Sustainable and Resilient Infrastructure,* **1**(3–4), 93–94.

Electric Power Research Institute. 1991. *A Methodology for Assessment of Nuclear Power Plant Seismic Margin*. Electric Power Research Institute, Palo Alto, CA.

Favier P, Poulos A, Vásquez J A et al. 2019. Seismic risk assessment of an emergency department of a Chilean hospital using a patient-oriented performance model, *Earthquake Spectra,* **35**(2), 489–512.

FEMA P695. 2009. *Quantification of Building Seismic Performance Factors*. FEMA, Washington DC.

FEMA P58. 2012. *Seismic Performance Assessment of Buildings*: Vols. 1 and 2–*Methodology*. Applied Technology Council (ATC), Redwood City, CA.

Filiatrault A, Sullivan T. 2014. Performance-based seismic design of nonstructural building components: The next frontier of earthquake engineering, *Earthquake Engineering and Engineering Vibration,* **13**(1), 17–46.

GB 50011-2010. 2010. *Code for Seismic Design of Buildings.* China Architecture and Building Press, Beijing (in Chinese).

Gilbert S, Ayyub B M. 2016. Models for the economics of resilience, *ASCE-ASME Journal of Risk and Uncertainty in Engineering Systems Part A*: *Civil Engineering*, **2**(4), 4016003.

Gul M, Guneri A F. 2015. A discrete event simulation model of an emergency department network for earthquake conditions. In: *6th International Conference on Modeling Simulation Applied Optimization (ICMSAO)*, Istanbul, Turkey, May 27–29, pp. 1–6.

Gunal M M. 2012. A guide for building hospital simulation models, *Health Systems,* **1**(1), 17–25.

Gunn S W A. 1995. Health effects of earthquakes, *Disaster Prevention & Management,* **4**(5), 6–10.

HAZUS-MH MR3. 2003. *Multi-Hazard Loss Estimation Methodology-Earthquake Model*, Technical Manual. Department of Homeland Security, Washington, DC.

Hassan E M, Mahmoud H. 2018. A framework for estimating immediate interdependent functionality reduction of a steel hospital following a seismic event, *Engineering Structures,* **168**, 669–683.

Hassan E M, Mahmoud H. 2019. Full functionality and recovery assessment framework for a hospital subjected to a scenario earthquake event, *Engineering Structures,* **188**, 165–177.

Hu Y, Dargush G F, Shao X. 2012. A conceptual evolutionary aseismic decision support framework for hospitals, *Earthquake Engineering and Engineering Vibration,* **11**(4), 499–512.

Inter-Agency Standing Committee (IASC). 2005. Humanitarian health cluster: Pakistan earthquake, *Consolidated Health Situation Bulletin,* p. **2**, 27 October 2005.

Jacques C C, Mcintosh J, Giovinazzi S et al. 2014. Resilience of the Canterbury hospital system to the 2011 Christchurch earthquake, *Earthquake Spectra,* **30**(1), 533–554.

Kammouh O, Dervishaj G, Cimellaro G P. 2017. Quantitative framework to assess resilience and risk at the country level, *ASCE-ASME Journal of Risk and Uncertainty in Engineering Systems Part A*: *Civil Engineering,* **4**(1), 4017033.

Kammouh O, Cimellaro G P, Mahin S A. 2018. Downtime estimation and analysis of lifelines after an earthquake, *Engineering Structures,* **173**, 393–403.

Kammouh O, Zamani Noori A, Cimellaro G P, Mahin S A. 2019. Resilience assessment of urban communities, *ASCE-ASME Journal of Risk and Uncertainty in Engineering Systems Part A*: *Civil Engineering,* **5**(1), 4019002.

Kircher C A, Whitman R V, Holmes W T. 2006. HAZUS earthquake loss estimation methods, *Natural Hazards Review,* **7**(2), 45–59.

Kirsch T D, Mitrani-Reiser J, Bissell R, Sauer L M, Mahoney M, Holmes W T, de la Maza F. 2010. Impact on hospital functions following the 2010 Chilean earthquake, *Disaster Medicine and Public Health Preparedness,* **4**(2), 122–128.

Li J C, Wang T, Shang Q X. 2019. Probability-based seismic reliability assessment method for substation systems, *Earthquake Engineering and Structural Dynamics,* **48**, 328–346.

Liu X, Ferrario E, Zio E. 2017. Resilience analysis framework for interconnected critical infrastructures, *ASCE-ASME Journal of Risk and Uncertainty in Engineering Systems Part B*: *Mechanical Engineering,* **3**(2), 21001.

Lupoi A, Cavalieri F, Franchin P. 2012. Probabilistic seismic assessment of health-care systems at regional scale. In: *15th World Conference on Earthquake Engineering* Lisbon Portugal.

Masi A, Santarsiero G, Chiauzzi L. 2012. Vulnerability assessment and seismic risk reduction strategies of hospitals in Basilicata region (Italy). In: *15th World Conference on Earthquake Engineering* Lisbon Portugal.

McCabe O L, Barnett D J, Taylor H G et al. 2010. Ready willing able: A framework for improving the public health emergency preparedness system, *Disaster Medicine and Public Health Preparedness,* **4**(2), 161–168.

Mieler M, Stojadinovic B, Budnitz R J et al. 2013. *Toward Resilient Communities: A Performance-Based Engineering Framework for Design and Evaluation of the Built Environment*, PEER Report 2013/19. Pacific Earthquake Engineering Research Center University of California, Berkeley.

Mieler M, Stojadinovic B, Budnitz R et al. 2015. A framework for linking community-resilience goals to specific performance targets for the built environment, *Earthquake Spectra*. doi:10.1193/082213EQS237.

Miniati R, Iasio C. 2012. Methodology for rapid seismic risk assessment of health structures: Case study of the hospital system in Florence Italy, *International Journal of Disaster Risk Reduction,* **2**, 16–24.

Miniati R, Capone P, Hosser D. 2014. Decision support system for rapid seismic risk mitigation of hospital systems. Comparison between models and countries, *International Journal of Disaster Risk Reduction,* **9**, 12–25.

Myrtle R C, Masri S F, Nigbor R L et al. 2005. Classification and prioritization of essential systems in hospitals under extreme events, *Earthquake Spectra,* **21**(3), 779–802.

Nakashima M, Lavan O, Kurata M, Luo Y. 2014. Earthquake engineering research needs in light of lessons learned from the 2011 Tohoku earthquake, *Earthquake Engineering and Engineering Vibration,* **13**(**S1**), 141–149.

Pali T, Macillo V, Terracciano M T. et al. 2018. In-plane quasi-static cyclic tests of nonstructural lightweight steel drywall partitions for seismic performance evaluation, *Earthquake Engineering & Structural Dynamics,* **47**(6), 1566–1588.

Paul J A, George S K, Yi P, Lin L. 2006. Transient modeling in simulation of hospital operations for emergency response, *Prehospital and Disaster Medicine,* **21**(4), 223–236.

Peng X, Roueche D B, Prevatt D O et al. 2016. An engineering-based approach to predict tornado-induced damage. In: *Multi-Hazard Approaches to Civil Infrastructure Engineering*. Springer, New York, pp. 311–335.

Porter K, Kennedy R, Bachman R. 2007. Creating fragility functions for performance-based earthquake engineering, *Earthquake Spectra,* **23**(2), 471–489.

Post Disaster Needs Assessments (PDNA). 2010. *Guidance for Health Sector Assessment to Support the Post Disaster Recovery Process*, Version 2.2. World Health Organization Humanitarian Health Action.

Schultz C H, Koenig K L, Lewis R J. 2003. Implications of hospital evacuation after the Northridge California earthquake, *New England Journal of Medicine,* **348**, 1349–1355.

Sørup C M, Jacobsen P, Forberg J L. 2013. Evaluation of emergency department performance: A systematic review on recommended performance and quality-in-care measures, *Scandinavian Journal of Trauma Resuscitation and Emergency Medicine,* **21**, 1–14.

Shang Q X, Wang T, Li J C. 2018a. Seismic fragility of flexible pipeline connections in a base isolated medical building, *Earthquake Engineering and Engineering Vibration*. doi:10.1139/cjce-2022-0222.

Shang Q X, Wang T, Li J C. 2018b. A quantitative framework to evaluate the seismic resilience of hospital systems, *Journal of Earthquake Engineering*. doi:10.1080/13632469.2020.1802371

Soroushian S, Rahmanishamsi E, Ryu K P et al. 2016. Experimental fragility analysis of suspension ceiling systems, *Earthquake Spectra,* **32**(2), 150504135037002.

Tavares D H, Padgett J E, Paultre P. 2012. Fragility curves of typical as-built highway bridges in eastern Canada, *Engineering Structures,* **40**(7), 107–118.

The MathWorks Inc. 2016. *Matlab Version R2016b*. The Math Works, Natick, MA.

United Nations Children's Fund (UNICEF). 2004. *Crisis Appeal Earthquake in Bam Iran*. https://www.unicef.org/emerg/files/Emergencies_Iran_Flash_Appeal_130104.pdf.

Wang X L. 2005. Seismic reliability analysis for functional utility system in hospital, Ph.D. Thesis. Beijing University of Technology (in Chinese).

Wang Y M, Xiong L H, Xu W X. 2013. Seismic damage and damage enlightenment of medical buildings in Lushan *MS* 7.0 earthquake, *Earthquake Engineering and Engineering Dynamics,* **4**(4), 44–53 (in Chinese).

World Health Organization (WHO). 2006. *Health Facility Seismic Vulnerability Evaluation: A Handbook, Outlined the Structural Vulnerability Function.* WHO Regional Office for Europe DK-2100, Copenhagen, Denmark.

Yavari S, Chang S E, Elwood K J. 2010. Modeling post-earthquake functionality of regional healthcare facilities, *Earthquake Spectra,* **26**(3), 869–892.

V'asquez A, Rivera F, de la Llera J, Mitrani-Reiser J. 2017. Healthcare network operation in Iquique after the 2014, Pisagua earthquake. In: *16th World Conference on Earthquake* 9–13 January 2017, Santiago Chile.

Vickery P J, Lin J, Skerlj P F. et al. 2006. HAZUS-MH hurricane model methodology. I: Hurricane hazard terrain wind load modeling, *Natural Hazards Review,* **7**(2), 82–93.

Welch S J, Asplin B R, Stone-Griffith S, Davidson S J, Augustine J, Schuur J. 2011. Emergency department operational metrics measures and definitions: results of the second performance measures and benchmarking summit, *Annals of Emergency Medicine,* **58**(1), 33–40.

World Health Organization (WHO). 2007. Risk reduction in the health sector and status of progress. In: *Proceedings of Disaster Risk Reduction in the Healthcare Sector—Thematic Workshop.* World Health Organization (WHO). Geneva Switzerland.

Zimmerman R. 2004. Decision-making and the vulnerability of interdependent critical infrastructure. *Proceedings of the 2004 IEEE International Conference on Systems Man and Cybernetics.* Hague, Netherlands, October 10-13, pp. 4059–4063.

7 Infrastructure Systems and Healthcare Network in a Benchmark City

7.1. INTRODUCTION

Seismic resilience, which is defined as the ability of an engineering system (e.g., buildings, bridges, communities, and cities) to resist, recover from, and adapt to an earthquake (Bruneau et al. 2003), has attracted significant attention from academia and industry recently. The recovery capacity of a system is quantified as the variation of the functionality over time. The ability of the system to maintain functionality after earthquakes and the ability of the system to recover from earthquakes are two crucial properties that a system should possess before or after the occurrence of earthquakes (Yodo and Wang 2016). The concept of seismic resilience provides a new means of thinking about how to survive earthquakes and recovery.

An assessment of the seismic resilience of a complicated engineering system provides an understanding of the impact of earthquakes in terms of functionality degeneration (robustness), substitutable components (redundancy), recovery time and speed (rapidity), and the available resources (resourcefulness) of the system (Bruneau et al. 2003). The assessment results can help decision-makers to formulate effective strategies in all phases of the earthquakes (Cimellaro et al. 2010). Different frameworks for seismic resilience assessment of engineering systems have been proposed during the last two decades, and qualitative and quantitative methods have been used. Bruneau et al. (2003) developed a conceptual assessment framework and a system diagram to improve system resilience by system assessment and modification during the pre- and post-earthquake periods. Chang and Shinozuka (2004) proposed a quantification framework that relates the expected losses in future earthquakes to a community's seismic performance. Miles and Chang (2006) presented a comprehensive conceptual model of functionality recovery that compares the disparity between systems with different levels of disaster preparedness and mitigation decisions. Hu et al. (2012) proposed a conceptual evolutionary framework for aseismic decision support for hospitals to integrate a range of engineering and sociotechnical models. Bruneau and Reinhorn (2007) used the percentage of the healthy population, the patient/day treatment capacity, and the repair cost to quantify time-variant resilience. The Multidisciplinary Center for Earthquake Engineering Research (MCEER) has identified the key steps for quantifying the technical and organizational aspects of resilience. The uncertainties related to the intensity measures, response parameters, performance threshold, performance measures, losses, and recovery time are part of the MCEER framework (Cimellaro et al.

 DOI: 10.1201/9781003457459-7

2009). Uncertainties induced by these factors should be carefully considered in resilience assessment. A large number of studies have been conducted to include uncertainties in resilience assessment, for instance, the MCEER framework (Cimellaro et al. 2009), probability-based methods (Dong and Frangopol 2016; Li et al. 2019), and probabilistic resilience assessment frameworks (Burton et al. 2017; Shang et al. 2020). The MCEER framework was subsequently adopted by Cimellaro et al. (2011) to estimate the resilience of hospital systems using a meta-model. The interaction between the technical and organizational aspects was considered by using penalty factors. Cimellaro et al. (2010) proposed a framework that integrates direct and indirect losses and a model for the recovery of organizational efficiency to quantify the resilience of hospitals. Cimellaro et al. (2016) developed the PEOPLES framework for measuring resilience at different scales (e.g., individual building, city, region, and state); seven different dimensions of resilience were used in a layered approach. The San Francisco Bay Area Planning and Urban Research Association (SPUR) developed a framework for improving San Francisco's resilience through seismic mitigation policies. The expected seismic performance of buildings and lifeline systems during and after earthquakes are defined in the SPUR framework (SPUR 2008). The National Institute of Standards and Technology (NIST) developed the Community Resilience Planning Guide for Buildings and Infrastructure Systems (Guide) to help communities enhance resilience by incorporating short- and long-term measures and consider community social goals and their dependencies on the built environment and infrastructure systems (NIST 2016). The SYNER-G project proposed an integrated general framework for vulnerability assessment of the physical and socio-economic impact and losses of an earthquake and applied this framework to cities including Thessaloniki, L'Aquila, and Vienna (Kyriazis 2013; Pitilakis et al. 2014). At the national level, the Hyogo Framework for Action (HFA) (UNISDR 2007; Djalante et al. 2012) was created by the United Nations International Strategy for Disaster Reduction (UNISDR) to enable systematic planning, implementation, and evaluation of disaster risk reduction activities to ensure the resilience of nations and communities. The Sendai Framework for Disaster Risk Reduction (SFDRR) (UNISDR 2015; Kelman and Glantz 2015) was established based on the HFA to achieve a substantial reduction in disaster risk and losses over the next 15 years. Since 2018, FEMA has been working to develop a new pre-disaster hazard mitigation program, the Building Resilient Infrastructure and Communities (BRIC) program, to improve disaster preparedness, mitigation, response, and recovery programs and outcomes (FEMA 2020). More and more attention has been paid to resilience assessment and resilient society.

In the aforementioned frameworks, the scale of the target systems ranges from individual buildings and interconnected networks to the community and city levels. However, the resilience assessment results of different frameworks require calibration based on a benchmark model to ensure that the method is valid and can be used by decision-makers. One example of an existing benchmark model is the Centerville virtual community (CVC) developed by Ellingwood et al. (2016), which is a city of approximately 50,000. At the city level, the Mid-America Earthquake (MAE) Center developed the Memphis Testbed (MTB) to study seismic effects on the city of Memphis (Steelman and Hajjar 2008). Similarly, Noori et al. (2017)

developed a virtual city (VC-TI) based on the buildings in the city of Turin, Italy. The virtual city covers an area of 120.1 km², with a population of 850,000. The above-mentioned models can be used for seismic resilience assessment and calibration. However, the size of the CVC is relatively small. The VC-TI and MTB are representations of a typical European city and an American city that differ substantially from Chinese cities. Therefore, a benchmark city based on a medium-sized city located in the southeastern region of China is developed in this study. Nearly authentic data will be provided, and the benchmark model can be used to calibrate resilience assessment frameworks, evaluate the effects of different strategies for resilience improvement, and facilitate the construction of a seismic resilient city in China. The data format and hierarchy structure can also be extended to other benchmark cities around the world.

The benchmark model is based on Geographic Information System (GIS) models and can be used for secondary development. The benchmark city consists of a residential zone, business zone, industrial park, government agencies, schools, hospitals, and physical lifeline systems. The demographics, site conditions, and potential hazards of the benchmark city model were described in this study. Detailed information on the building inventory and different lifeline systems is also provided. Data from past earthquakes and the literature are used to develop fragility models, consequence models, and recovery models that can be utilized in the resilience assessment process. A case study on the accessibility of emergency rescue after earthquakes was conducted to demonstrate the completeness of the data included in the benchmark city. The ultimate goal is to provide a platform for calibrating resilience assessment results and help decision-makers to formulate effective strategies in all phases of the earthquakes, i.e., the planning for and the recovery from disasters.

7.2   DESCRIPTION OF THE BENCHMARK CITY

7.2.1   DEMOGRAPHICS

The benchmark model represents a medium-sized city in the southeastern coastal region of China. A summary of the demographics of the benchmark city is presented in Table 7.1. The total area of the benchmark city is 344.56 km². The urbanization level is typically represented by the proportion of the urban population to the total population and is 60% for this city. The Engel coefficient of urban residents is 33%, and the Gini coefficient is 0.3. The average number of years of education is 14 years, indicating that most of the citizens are high school graduates (12 years) or have a college diploma (≥16 years). The unemployment rate of the urban residents is 2.10%.

7.2.2   EARTHQUAKE HAZARD

The design earthquake intensity is VII, and the peak ground acceleration (PGA) corresponds to a design-basis earthquake (DBE) is 0.10 g with an exceedance probability of 10% in 50 years. The characteristic period T_g of the site of the entire benchmark city is 0.40 s, according to GB 18306-2015 (2015). The design earthquake group for this benchmark city is categorized as the second group in the Chinese code

TABLE 7.1
Population Demographics of the Benchmark City

Indicator	Value
Total area (km²)	344.56
GDP per capita ($)	16,218
Urbanization level	60%
Disposable income of urban residents ($)	11,141
Per capita net income of rural residents ($)	6,346
Engel coefficient of urban residents	33%
Gini coefficient	0.3
Average number of years of education (year)	14
Unemployment rate of urban residents	2.10%

(GB 50011-2010 2010). Existing ground-motion models for Chinese cities, such as those provided by Hong and Feng (2019), can be used to map the seismic hazard of this benchmark city.

7.2.3 LAND USE OF RESIDENTIAL AREAS

The residential area consists of five city zones and 14 residential areas, as shown in Figure 7.1. The number of citizens in the 14 residential areas is shown in Figure 7.1. The residential areas are composed of 337 residential units. The business district covers the center, western, southern, and northern parts of the city. The government center and central business district (CBD) are located in the southern part of the city. Educational areas are scattered all over the city. The northwestern part of the residential area (North district) is an industrial area. The population of the benchmark city is about 690,000. The 346 residential units are shown in Figure 7.2.

7.2.4 BUILDING INVENTORY

The buildings in the benchmark city have several construction types, as shown in Figure 7.3a. There are 9,773 buildings in the benchmark city, including 946 unreinforced masonry structures, 3,778 reinforced concrete (RC) frame structures, 1,621 RC frame shear wall structures, 3,420 reinforced masonry structures, and eight steel frame structures. The proportions of the different structural types are shown in Figure 7.3c. The vast majority of buildings are low-rise buildings (three stories or less, 38.45%) and multi-story buildings (four stories to six stories, 36.75%). These buildings were constructed in different years and were thus designed following different design codes. However, most of the buildings were built after 1989, accounting for 89.56% of the total. During the past three decades, the seismic design codes in China have changed three times, namely, GBJ 11–89 (1989), GB 50011-2001 (2001), and GB 50011-2010 (2010). The proportions of buildings constructed in different years are shown in Figure 7.3b. The total occupied area of these buildings is 5.558 km², and the total building area covering all building floors is 34.966 km². The

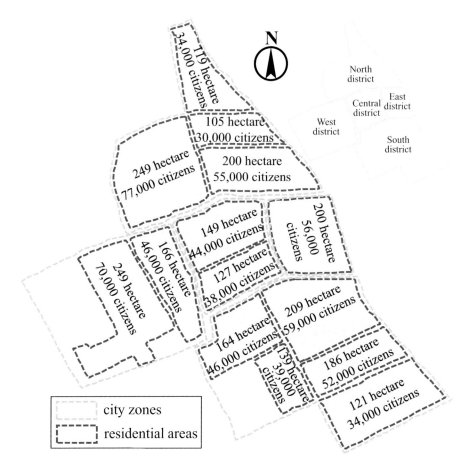

FIGURE 7.1 Five city zones and 14 residential areas of the benchmark city.

proportions of occupied areas and building areas of different types of buildings are shown in Figure 7.3d and 7.3e. All the unreinforced masonry structures were built before 1989, and 81.8% are one-story buildings. The statistics of the building height and the building numbers are provided in Figure 7.3f– h. It is worth noting that there are eight super high-rise buildings in this city, and their heights (100 m–256 m) are not listed in Figure 7.3h.

7.3 INFRASTRUCTURE SYSTEMS

Interdependent systems, including individual buildings and lifeline systems in a city, constitute a complex network system. Interdependencies between these systems often exacerbate the initial damage, leading to cascading failures during earthquakes and causing problems for emergency rescues after earthquakes. However, most currently available frameworks or methods focus primarily on one system and do not address

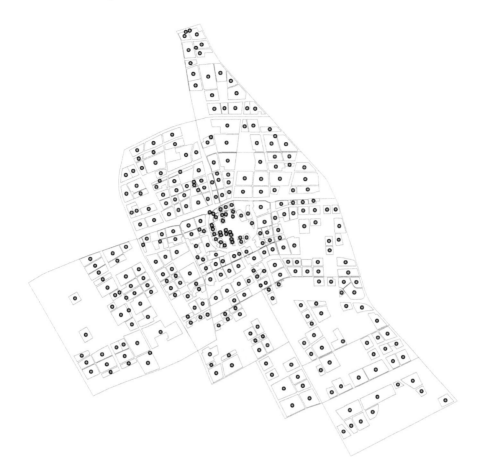

FIGURE 7.2   Distribution of residential units.

the importance of assessing interdependencies (Reiner and McElvaney 2017). For instance, building damages will result in injuries and death after a huge earth-quake. The delivery of emergency rescue instructions relies on the communication system while its operationality depends on the power distribution network. The damage of the transportation system will influence the delivery speed of injured people to hospitals. The medical treatments in hospitals need electricity to ensure the normal operation of medical equipment. Based on the building inventory infor-mation in Section 7.2 and detailed information on lifeline systems in this section, the interdependencies can be considered in seismic resilience assessments. Lifeline systems of the developed benchmark city are designed based on Chinese codes and the city planning documents (PAR 2017). Basic information on the lifeline systems, including the transportation system, the power, water, drainage, and natural gas net-works, as well as hospitals, emergency shelters, and schools, will be discussed later.

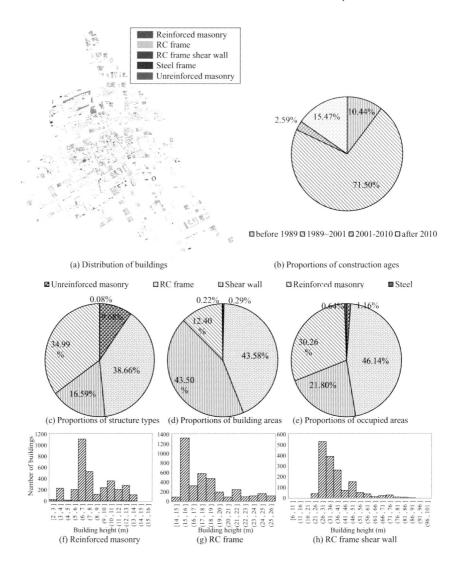

FIGURE 7.3   Building inventory information.

Resilience assessments of individual systems or the entire city and its interconnected and interdependent lifeline systems are feasible using this benchmark model.

7.3.1 Power Distribution Network

The power distribution network is shown in Figure 7.4. It includes fourteen 110 kV substations, four 220 kV substations, and the distribution lines. The substations are used for the power supply of the benchmark city or to connect to the power grid of other cities or regions. It is worth noting that "outside" represents substations located outside the benchmark city; these are not considered in the model. The capacity of the power distribution network is listed in Table 7.2.

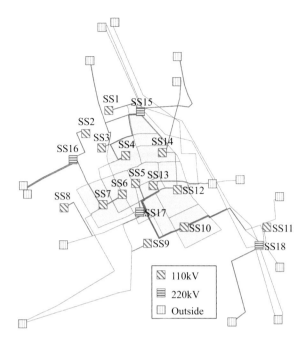

FIGURE 7.4   Power distribution network.

TABLE 7.2
Capacity Information of the Power Distribution Network

Substation Name	Voltage (kV)	Capacity (MVA)	Substation Name	Voltage (kV)	Capacity (MVA)
SS1	110	3×80	SS10	110	3×80
SS2	110	3×80	SS11	110	2×80
SS3	110	3×80	SS12	110	3×80
SS4	110	2×80	SS13	110	2×80
SS5	110	2×80	SS14	110	2×80
SS6	110	3×80	SS15	220	3×240
SS7	110	3×80	SS16	220	3×240
SS8	110	3×80	SS17	220	2×240
SS9	110	3×80	SS18	220	3×240

7.3.2 TRANSPORTATION SYSTEM

The transportation system of the benchmark city consists of roads of different levels with different travel speeds under normal conditions. High-speed roads, major roads, subsidiary roads, and small branch roads comprise the road system. The transportation system also includes 10 masonry arch bridges and 45 simply-supported girder bridges, as shown in Figure 7.5a. The recommended travel speed ranges from 20 to 70 km/h, as shown in Figure 7.5b. The total number of roads is 1,133, with a total

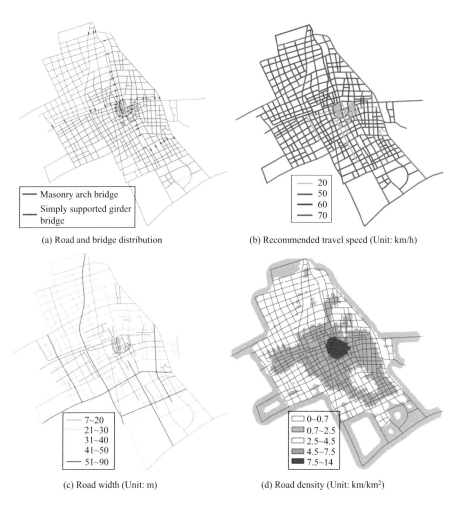

(a) Road and bridge distribution

Masonry arch bridge
Simply supported girder bridge

(b) Recommended travel speed (Unit: km/h)

— 20
— 50
— 60
— 70

(c) Road width (Unit: m)

— 7~20
— 21~30
31~40
41~50
— 51~90

(d) Road density (Unit: km/km²)

☐ 0~0.7
☐ 0.7~2.5
☐ 2.5~4.5
■ 4.5~7.5
■ 7.5~14

FIGURE 7.5 Transportation system.

TABLE 7.3
Road Hierarchy and Recommended Travel Speed

Road Hierarchy	Recommended Travel Speed (km/h)	Road Length (m)
High-speed road	70	46,684
Arterial road	60	149,518
Secondary road	50	252,209
Branch road	20	37,263

length of 485.7 km. The recommended travel speeds of the different road levels are listed in Table 7.3, and the road width and total road length are listed in Table 7.4. The transportation network, including the road width and the density of the road network, are shown in Figure 7.5c and d.

TABLE 7.4
Road Width and Road Length

Road Width (m)	Road Length (m)
7–20	93,075
21–30	159,665
31–40	94,174
41–50	89,297
51–90	49,464

7.3.3 Water Distribution Network

The urban water consumption in the benchmark city is 219,164 m³/day. The benchmark city is supplied with water by a water plant outside the city (with an output of 600,000 m³/day). Water to the citizens is supplied by two pressure-boosting stations (PBSs), namely, PBS 1 (with an output of 200,000 m³/day) and PBS 2 (with an output of 50,000 m³/day). The water plant (WP 1) (Figure 7.6) is used as an emergency standby water plant (with an output of 150,000 m³/day). The water distribution network consists of 323 pipe segments, with a total length of 255 km. The diameter, length, and material of the pipe segments, as well as the water demand of the nodes under normal conditions, are provided in the benchmark model (Table 7.5). The water distribution network is located in a flood plain; thus, the elevations of the nodes are zero for simplicity.

7.3.4 Drainage Network

The volume of the centralized sewage treatment plant in the benchmark city is about 163,000 m³/day. The sewage treatment volumes of the four sewage plants (SPs), namely, SP1, SP2, SP3, and SP4 are 20,000, 80,000, 30,000, and 80,000 m³/day, respectively. The sewage treatment volumes of the 18 pump stations are listed in Table 7.6. The drainage network is shown in Figure 7.7. The pipe diameter and length information of the drainage network are provided in Table 7.7; the diameters of the drainage pipes range from DN400 to DN1200.

7.3.5 Natural Gas Distribution Network

The natural gas distribution network serves the residential units. The total gas consumption is about 970 million m³/year. As shown in Figure 7.8, the gas gate station and regulator stations receive gas from a high-pressure pipe from the supply source outside the city. The gas is sent to the medium-pressure natural gas pipe of each residential area after filtration, metering, and pressure adjustment. The liquid natural gas (LNG) supply station is used as a peak regulator gas source or emergency standby gas source. The pipe diameter and length information of the natural gas distribution network are listed in Table 7.8.

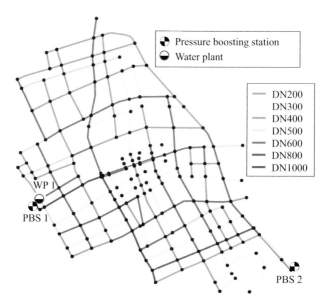

FIGURE 7.6   Water distribution network.

TABLE 7.5
Details of the Pipes Used in the Water Distribution Network

Diameter (mm)	Material	Number	Total Length (m)	Material	Number	Total Length (m)
200	Cast iron	7	6,340	PVC	0	0
300	Cast iron	60	39,285	PVC	25	22,506
400	Cast iron	33	26,337	PVC	57	51,265
500	Cast iron	8	6,476	PVC	14	11,783
600	Cast iron	62	45,814	PVC	23	22,015
800	Cast iron	11	8,545	PVC	0	0
1,000	Cast iron	22	13,703	PVC	1	605
Sum	–	323	254,674			

7.3.6 Hospitals, Emergency Shelters, and Schools

There are eight hospitals in the benchmark city, including four Level 2 hospitals and four Level 3 hospitals of different sizes (Table 7.9). The construction types of the hospitals include RC frames and RC frame shear wall systems. A total number of 4,657 patient beds and 4,995 medical staff are available in emergency situations during an earthquake. The numbers of clinical departments and medical laboratories in each hospital are listed in Table 7.9. The locations of the eight hospitals in the residential units are shown in Figure 7.9. In addition to the eight hospitals, there are 20 emergency shelters, which can be used for temporary medical service in case of an emergency. The locations of the educational institutions,

TABLE 7.6
Sewage Treatment Volumes of Pump Stations

Pump Station Name	Volume ($10^3 m^3$/day)	Pump Station Name	Volume ($10^3 m^3$/day)
JX1	10	J1	6
JX2	25	J2	20
JX3	40	J3	10
X1	10	J4	20
X2	25	N1	20
X3	40	N2	15
X4	18	N3	20
X5	18	N4	20
X6	60	N5	15

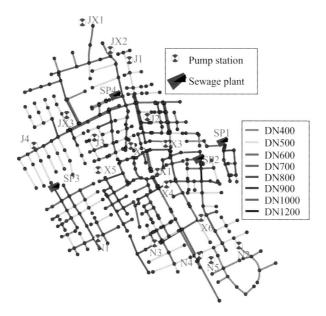

FIGURE 7.7   Drainage network.

including 15 primary schools, eight junior middle schools, and three senior middle schools, are shown in Figure 7.9. These schools can also serve as emergency evacuation sites.

7.4   BASIC MODELS FOR SEISMIC RESILIENCE ASSESSMENT

7.4.1   Seismic Fragility Models

A seismic fragility model is used to define the probability that a component, building, or system exceeds a pre-defined damage state (DS) based on a given engineering demand parameter (EDP), e.g., PGA, peak floor acceleration (PFA), and the

TABLE 7.7

Pipe Information of the Drainage Network

Pipe Diameter	Pipe Length (m)
DN400	27,988
DN500	62,045
DN600	47,069
DN700	1,560
DN800	45,547
DN900	4,361
DN1000	27,075
DN1200	1,099
Total length	216744

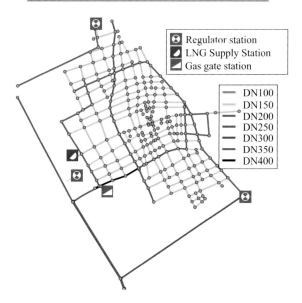

FIGURE 7.8   Natural gas distribution network.

maximum inter-story drift ratio (IDR). A fragility model is usually idealized as a lognormal distribution function and has been used in seismic resilience assessments to obtain probability-based results. The parameters (including the median values (x_m) and logarithmic standard deviations (β)) used to define the fragility curve with a lognormal distribution have a substantial influence on the assessment results and, therefore, should be carefully selected. Considering that the design codes in China are different from other countries, the fragility models provided by HAZUS-MH software package (HAZUS-MH MR3 2003) and FEMA P58 (2012) may be not suitable for the seismic performance of buildings and infrastructures in China. Therefore, experimental results, numerical results, and earthquake reconnaissance results from available literature are collected to develop fragility models that can be

TABLE 7.8

Pipe Information of the Natural Gas Distribution Network

Pipe Diameter	Pipe Length (m)
DN100	63,799
DN150	152,298
DN200	46,469
DN250	31,943
DN300	19,099
DN350	5,851
DN400	3709
Total length	323,169

TABLE 7.9

Detailed Information on the Eight Hospitals

Hospital Name	Hospital Level	Number of Stories	Patient Beds	Medical Staff	Clinical Departments	Medical Laboratories
H-N-1	Level 2	6	450	576	33	7
H-C-1	Level 3	6	511	731	28	13
H-C-2	Level 3	10	1,248	1,785	25	7
H-E-1	Level 2	5	150	192	10	4
H-W-1	Level 2	5	120	154	8	3
H-W-2	Level 2	5	218	279	23	7
H-W-3	Level 3	10	900	1,287	24	7
H-S-1	Level 3	10	1,000	1,430	28	8

used for the proposed benchmark models. Examples of the fragility parameters of different buildings and bridges are listed in Table 7.10.

The statistical data obtained by Taghavi and Miranda (2003) indicated that non-structural components (NSCs) accounted for most of the investment in a typical building. Although damage to structural components (SCs) is the most important measure of building damage affecting casualties, the damage to NSCs results in the highest economic loss and loss of functionality. Therefore, both SCs and NSCs should be considered in seismic performance or resilience assessments (Filiatrault and Sullivan 2014). The recent Chinese standard for the seismic resilience assessment of buildings (GB/T 38591-2020 2020) provides fragility database for NSCs in China and can be used for seismic resilience of the benchmark model. The fragility parameters of the NSCs can also be obtained from the literature on the seismic fragility of NSCs, such as piping systems (Wang et al. 2019), partition walls (Pali et al. 2018), and suspended ceilings (Lu et al. 2018). The fragility curves of drift-sensitive NSCs and acceleration-sensitive NSCs provided by the HAZUS-MH software package (HAZUS-MH MR3 2003) can also be used as a compromise if fragility parameters of some NSCs are not available.

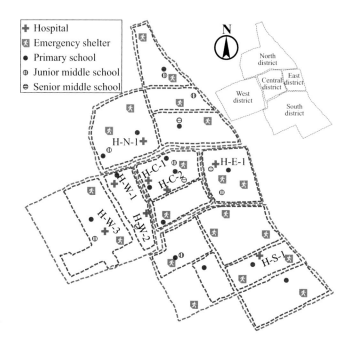

FIGURE 7.9 Hospitals, emergency shelters, and schools.

TABLE 7.10
Fragility Parameters of Different Buildings and Bridges

	EDP	Parameter	Slight Damage	Moderate Damage	Extensive Damage	Complete Damage	Reference
Unreinforced masonry	IDR	x_m	0.0008	–	0.002	0.0045	Jiang et al. (2020)
		β	0.35	–	0.30	0.35	
Reinforced masonry	IDR	x_m	0.00063	0.00143	0.00286	0.005	Xiong (2004)
		β	0.35	0.35	0.30	0.35	
RC frame	IDR	x_m	0.00182	0.01	0.02	0.04	Yu et al. (2016)
		β	0.20	0.30	0.30	0.40	
RC frame shear wall	IDR	x_m	0.00125	0.0025	0.005	0.01	Xu (2019)
		β	0.40	0.40	0.40	0.40	
Steel frame	IDR	x_m	0.00333	0.005	0.01	0.01818	Li et al. (2017)
		β	0.40	0.40	0.40	0.40	
Masonry arch bridge	PGA (g)	x_m	0.251	0.475	0.907	2.539	Lin et al. (2017)
		β	1.004	0.887	0.845	0.845	
Simply-supported girder bridge	PGA (g)	x_m	0.237	0.596	1.262	1.286	Lin et al. (2017)
		β	0.998	0.751	0.704	0.357	

7.4.2 Consequence Models

A strong correlation is assumed between building damage (both structural and non-structural damage) and the number of casualties and economic loss. The HAZUS-MH software package (HAZUS-MH MR3 2003) and FEMA P58 (2012) provide consequence models for evaluating the number of casualties and economic losses of buildings. For instance, HAZUS-MH MR3 (2003) defines the four levels of casualty severities as light injuries, hospitalized injuries, life-threatening injuries, and deaths and uses event trees to determine the number of casualties caused by an earthquake. FEMA P58 (2012) provides casualty rates for different SCs and NSCs in different damage states. Hingorania et al. (2020) summarized the relevant parameters for predicting potential fatalities and developed consequence models for predicting the loss of life due to the collapse of buildings. However, these data and models may be not suitable to be used in Chinese buildings. The Chinese code GB/T 18208.4-2011 (2012) provides models for evaluating direct economic losses caused by earthquakes. GB/T 38591-2020 (2020) states that the number of death should be calculated using the product of the personnel density, floor area, and death rate corresponding to each damage state. The number of injuries can be calculated in the same manner, with injury rates corresponding to each damage state. The criteria for determining the damage states of floors and the personnel density of buildings with different functions are provided in GB/T 38591-2020 (2020). The consequence models provided in GB/T 18208.4-2011 (2012) and GB/T 38591-2020 (2020) can be used for seismic resilience assessment of the developed benchmark model.

7.4.3 Recovery Models

FEMA P58 (2012) provides a repair sequence considering parallel repair and series repair for building level recovery. Based on the determined repair schedule, the recovery process of a damaged building can be evaluated in terms of the recovery time and functionality improvement. More detailed repair paths, such as the REDi (Almufti and Willford 2013) repair sequence and the repair path provided by Shang et al. (2020) can also be selected for building recovery assessment. For city-level building recovery, the repair schedule for regional buildings developed by Xiong et al. (2020) can be used. However, the recovery process of buildings after earthquakes is highly dependent on the lifeline systems. Therefore, the recovery process of lifeline systems is also a key factor in evaluating the seismic resilience of a city. Kammouh et al. (2018) developed recovery models based on statistical data of downtime for different lifeline systems in recent earthquakes. The models are presented in terms of the probability of recovery time. However, the statistical data were collected from different countries worldwide, and the recovery model may not be suitable for Chinese cities. Therefore, in this study, recovery time data of lifeline systems after recent earthquakes in China were analyzed to generate time-based recovery models, as depicted in Figure 7.10. The earthquake magnitude (EM) is used as the

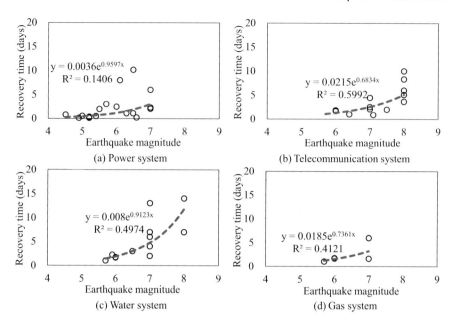

FIGURE 7.10 Recovery time of lifeline systems.

intensity measure since most datasets include this parameter. The recovery model describes the relationship between the recovery time of a system after earthquakes and the earthquake magnitude. Data with EM ranging from 4.5 to 8.0 were selected to develop the recovery models. It is observed that the recovery time increases slowly when EM is small and the increasing trend becomes quite large when EM is large. An exponential function (Equation 7.1) is adopted to describe the relationship between recovery time and EM considering both the distribution of data and the goodness of fitting. The proposed recovery models can be used to calibrate the evaluation results of different frameworks. It is noteworthy that the data from four past earthquakes in China were used to generate recovery time model for the gas system. The accuracy shall be noted because of the limited number of samples.

$$T = ae^{bx} \tag{7.1-1}$$

$$\ln(T) = \ln(a) + bx \tag{7.1-2}$$

The functionality of a system is improved when the recovery process has been initiated after earthquakes, and the change in functionality over time is usually described using a resilience curve. Figure 7.11 presents several examples of resilience curves of lifeline systems reported after recent earthquakes (Monsalve and Llera 2019). For most cases in Figure 7.11, the lifeline systems recover quickly after the earthquakes, and the functionality is improved rapidly. After the initial recovery, the rate of recovery decreases. It is observed that the recovery time of lifeline systems internationally

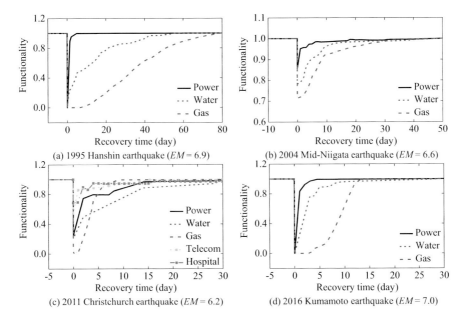

FIGURE 7.11   Resilience curves of lifeline systems in recent earthquakes (Monsalve and Llera 2019).

is substantially longer than that in China (compared with the predicted data in Figure 7.10). This result is attributed to differences in the political system and policies between China and other countries.

7.5   SUMMARY AND CONCLUSIONS

It is well-known that seismic resilience evaluations of cities require interdisciplinary expertise and numerous data models. Current evaluation results of different frameworks or methods have to be calibrated based on a unified evaluation model. To promote the seismic resilience assessment and calibration of different frameworks at a city level, this study presents a novel benchmark model of a medium-sized city located in the southeastern coastal region of China. The developed model can serve as a testbed to facilitate the construction of a seismic resilient city in China. Similar testbeds of American and European cities such as Centerville virtual community (CVC), Memphis Testbed (MTB), and virtual city of Turin, Italy (VC-TI) are available now. However, Chinese cities are quite different from American and European cities, and such testbed is urgently needed.

Detailed information on the benchmark city, including demographics, site conditions, potential hazard exposure, and building inventory is provided. Descriptions of lifeline systems including power, transportation, water, drainage, and natural gas distribution networks, as well as hospitals, emergency shelters, and schools designed based on Chinese codes, are also provided. The basic models including the seismic fragility models, consequence models, and recovery models are developed based on post-earthquake data obtained from the literature. These models are suitable for seismic resilience assessment and calibration. With these data and models, the seismic performance of the city can be quantified by available assessment frameworks.

As the first stage of the development of the benchmark city, this study bears some limitations and several extensions need to be explored in future studies. First, the recovery models developed based on limited data need to be updated by collecting more related data. Second, analyses of different systems still need to be conducted to comprehensively demonstrate the data completeness. Third, how to quantify the interdependency between different infrastructures and the associated uncertainties still needs more effort.

REFERENCES

Almufti I, Willford M. 2013. *REDiTM Rating System: Resilience-Based Earthquake Design Initiative for the Next Generation of Buildings.* Arup, San Francisco, CA.

Bruneau M, Chang S E, Eguchi R T, Lee G C, O'Rourke T D, Reinhorn A M, Shinozuka M, Tierney K, Wallace W A, Winterfeldt D V. 2003. A framework to quantitatively assess and enhance the seismic resilience of communities, *Earthquake Spectra*, **19**(4), 733–752.

Bruneau M, Reinhorn A M. 2007. Exploring the concept of seismic resilience for acute care facilities, *Earthquake Spectra*, **23**(1), 41–62.

Burton H V, Deierlein G, Lallemant D, Singh Y. 2017. Measuring the impact of enhanced building performance on the seismic resilience of a residential community, *Earthquake Spectra*, **33**(4), 1347–1367.

Chang S E, Shinozuka M. 2004. Measuring improvements in the disaster resilience of communities, *Earthquake Spectra*, **20**(3), 739–755.

Cimellaro G P, Fumo C, Reinhorn A M, Bruneau M. 2009. *Quantification of Disaster Resilience of Health Care Facilities*, Technical Report, MCEER-09-0009. MCEER.

Cimellaro G P, Reinhorn A M, Bruneau M. 2010. Seismic resilience of a hospital system, *Structure & Infrastructure Engineering*, **6(1–2)**, 127–144.

Cimellaro G P, Reinhorn A M, Bruneau M. 2011. Performance-based metamodel for healthcare facilities, *Earthquake Engineering & Structural Dynamics*, **40**(11), 1197–1217.

Cimellaro G P, Renschler C, Reinhorn A M, Arendt L. 2016. PEOPLES: A framework for evaluating resilience, *Journal of Structural Engineering*, **142**(10), 04016063.

Djalante R, Thomalla F, Sinapoy M S, Carnegie M. 2012. Building resilience to natural hazards in Indonesia: Progress and challenges in implementing the Hyogo Framework for Action, *Natural Hazards*, **62**(3), 779–803.

Dong Y, Frangopol D M. 2016. Probabilistic assessment of an interdependent healthcare-bridge network system under seismic hazard, *Structure and Infrastructure Engineering*, **13**(1), 160–170.

Ellingwood B R, Cutler H, Gardoni P, Peacock W G, Van d L J W, Wang NY. 2016. The centerville virtual community: A fully integrated decision model of interacting physical and social infrastructure systems, *Sustainable & Resilient Infrastructure*, **1**(3–4), 95–107.

FEMA P58. 2012. *Seismic Performance Assessment of Buildings,* Vol. 1: *Methodology.* Applied Technology Council (ATC), Redwood City, CA.

FEMA. 2020. *Summary of Stakeholder Feedback: Building Resilient Infrastructure and Communities (BRIC).* Federal Emergency Management Agency, Washington, DC

Filiatrault A, Sullivan T. 2014. Performance-based seismic design of nonstructural building components: The next frontier of earthquake engineering, *Earthquake Engineering and Engineering Vibration*, **13**(1), 17–46.

GBJ 11-89. 1989. *Code for Seismic Design of Buildings*. China Architecture and Building Press, Beijing (in Chinese).

GB 50011-2001. 2001. *Code for Seismic Design of Buildings*. China Architecture and Building Press, Beijing (in Chinese).

GB 50011-2010. 2010. *Code for Seismic Design of Buildings*. China Architecture and Building Press, Beijing (in Chinese).

GB 18306-2015. 2015. *Seismic Ground Motion Parameters Zonation Map of China*. Standardization Administration of China, Beijing (in Chinese).

GB/T 18208.4-2011. 2012. *Post-Earthquake Field Works—Part 4: Assessment of Direct Loss*. Standardization administration of China, Beijing (in Chinese).

GB/T 38591-2020. 2020. *Standard for Seismic Resilience Assessment of Buildings*. Standardization Administration of China, Beijing (in Chinese).

Gunn S W A. 1995. Health effects of earthquakes, *Disaster Prevention & Management*, **4**(5), 6–10.

HAZUS-MH MR3. 2003. *Multi-Hazard Loss Estimation Methodology-Earthquake Model, Technical Manual*. Department of Homeland Security, Washington, DC.

Hingorania R, Tannera P, Prietob M, Laraa C. 2020. Consequence classes and associated models for predicting loss of life in collapse of building structures, *Structural Safety*, **85**, 101910.

Hong H P, Chao F. 2019. On the ground-motion models for chinese seismic hazard mapping, *Bulletin of the Seismological Society of America*, **109**(5), 2106–2124.

Hu Y, Dargush G F, Shao XY. 2012. A conceptual evolutionary aseismic decision support framework for hospitals, *Earthquake Engineering and Engineering Vibration*, **11**(4), 499–512.

Jiang YP, Su L and Huang X. 2018. Seismic fragility analysis of unreinforced masonry structures considering parameter uncertainties, *Engineering Mechanics*, **37**(1), 159–167 (in Chinese).

Kammouh O, Cimellaro G P, Mahin S A. 2018. Downtime estimation and analysis of lifelines after an earthquake, *Engineering Structures*, **173**: 393–403.

Kelman I, Glantz M H. 2015. Analyzing the Sendai framework for disaster risk reduction, *International Journal of Disaster Risk Science*, **6**(2), 105.

Kyriazis P. 2013. *Syner-G Final Report: Systemic Seismic Vulnerability and Risk Analysis for Buildings, Lifeline Networks and Infrastructures Safety Gain*, CP-IP 244061. Aristotle University of Thessaloniki, Thessaloniki, Greece.

Li Y M, Li Y Z, Yang B Y. 2017. Performance-based seismic fragility analysis of steel frame structures, *Earthquake Resistant Engineering and Retrofitting*, **39**(4), 55–59 (in Chinese).

Li JC, Wang T, Shang Q X. 2019. Probability-based seismic reliability assessment method for substation systems, *Earthquake Engineering and Structural Dynamics*, **48**: 328–346.

Lin Q L, Lin J Q, Liu J L. 2017. A study on the fragility of highway bridges in the Wenchuan earthquake, *Journal of Vibration and Shock*, **36**(4), 110–118 (in Chinese).

Lu Y, Mosqueda G., Han Q, Zhao Y. 2018. Shaking table tests examining seismic response of suspended ceilings attached to large-span spatial structures, *Journal of Structural Engineering*, **144**(9), 04018152.

Miles S B, Chang S E. 2006. Modeling community recovery from earthquakes, *Earthquake Spectra*, **22**(2), 439–458.

Monsalve M, de la Llera J C. 2019. Data-driven estimation of interdependencies and restoration of infrastructure systems, *Reliability Engineering & System Safety*, **181**: 167–180.

National Institute of Standards and Technology (NIST). 2016. *Community Resilience Planning Guide for Buildings and Infrastructure Systems*, NIST Special Publication 1190, U.S. Department of Commerce, Washington, DC.

Nie G Z, Gao J G, Su G W, Wang J M. 2001. Models on rapid judgement for the emergent rescue needs during earthquake, **23**(1), 69–76 (in Chinese).

Noori A Z, Marasco S, Kammouh O, Domaneschi M, Cimellaro G P. 2017. Smart cities to improve resilience of communities. In: *8th International Conference on Structural Health Monitoring of Intelligent Infrastructure*, Brisbane, Australia, December 5–8.

Pali T, Macillo V, Terracciano M T, Bucciero B, Fiorino L, Landolfo R. 2018. In-plane quasi-static cyclic tests of nonstructural lightweight steel drywall partitions for seismic performance evaluation, *Earthquake Engineering & Structural Dynamics*, **47**(6), 1566–1588.

Planning Administration of Rugao (PAR). 2017. *General Urban Planning of Rugao (2013-2030)*. Planning Administration of Rugao, Rugao (in Chinese).

Pitilakis K, Franchin P, Khazai B, Wenzel H. 2014. *SYNER-G: Systemic Seismic Vulnerability and Risk Assessment of Complex Urban, Utility, Lifeline Systems and Critical Facilities: Methodology and Applications*, vol. **31**. Springer, Dordrech.

Reiner M, McElvaney L. 2017. Foundational infrastructure framework for city resilience, *Sustainable and Resilient Infrastructure*, **2**(1), 1 7.

San Francisco Bay Area Planning and Urban Research Association (SPUR). 2008. *The Resilient City: Defining What San Francisco Needs From its Seismic Mitigation Policies*, SPUR Report, SPUR Board of Directors.

Shang Q X, Wang T, Li J C. 2020. Seismic resilience assessment of emergency departments based on the state tree method, *Structural Safety*, **85**, 101944.

Steelman J S, Hajjar J F. 2008. *Capstone Scenario Applications of Consequence-Based Risk Management for the Memphis Testbed*. Mid-American Earthquake Center, Department of Civil and Environmental Engineering, University of Illinois at Urbana-Champaign, Urbana, IL.

Taghavi S, Miranda E. 2003. *Response Assessment of Nonstructural Building Elements*, PEER Report 2003/05. Pacific Earthquake Engineering Research Center, Berkeley, CA.

United Nations International Strategy for Disaster Reduction (UNISDR). 2007. *Words Into Action: A Guidance for Implementing the Hyogo framework—Hyogo Framework for Action 2005–2015: Building the Resilience of Nations and Communities to Disasters*. UNISDR, Geneva Switzerland.

United Nations International Strategy for Disaster Reduction (UNISDR). 2015. *Sendai Framework for Disaster Risk Reduction 2015–2030*. UNISDR, Geneva Switzerland.

Wang T, Shang Q X, Chen X, Li J C. 2019. Experiments and fragility analyses of piping systems connected by grooved fit joints with large deformability, *Frontiers in Built Environment*, **5**, 49.

Xiong L H. 2004. Study on seismic performance of small hollow concrete block buildings, Ph.D. Dissertation. Institute of Engineering Mechanics, China Earthquake Administration (in Chinese).

Xiong C, Huang J, Lu X Z. 2020. Framework for city-scale building seismic resilience simulation and repair scheduling with labor constraints driven by time-history analysis, *Computer-Aided Civil and Infrastructure Engineering*, **35**(4), 322–341.

Xu M Y. 2019. Seismic fragility analysis of frame-shear structures based on new intensity measures and CPU parallel computing, Master Dissertation. Harbin Institute of Technology, Harbin (in Chinese).

Yodo N, Wang P. 2016. Engineering resilience quantification and system design implications: A literature survey, *Journal of Mechanical Design*, **138**(11), 111408.

Yu X H, Lu D, Li B. 2016. Estimating uncertainty in limit state capacities for reinforced concrete frame structures through pushover analysis, *Earthquakes and Structures*, **10**(1), 141–161.

8 Post-Earthquake Healthcare Service Accessibility in a Benchmark City

8.1 INTRODUCTION

Public service facilities such as schools, hospitals, and shopping centers provide basic services to citizens in modern cities. Healthcare service is provided primarily by hospitals. For instance, secondary and tertiary hospitals in China provide critical services for ensuring public health. The World Health Organization (WHO) strongly advocates for health equalities between different people and regions within a country (WHO 2005). Disparities in accessibility to healthcare service have received growing attention as a public concern over recent decades.

Accessibility reveals the relative ease for citizens to reach a service (e.g., healthcare) from a given location. Accessibility is inseparable from the starting point (origin), the end point (destination), and the connection between them (transportation network). These three elements together constitute "accessibility" by definition. Different methods for measuring accessibility to healthcare service have been proposed, including provider-to-population ratio (Guagliardo 2004), travel impedance (Dutt et al. 1986), gravity model (Hansen 1959), two-step floating catchment area (2SFCA) (Radke & Mu 2000; Luo & Wang 2003), and Kernel density (Spencer & Angeles 2007) methods.

Accessibility assessment results can be used to determine the equity of healthcare service within cities and help relevant decision-makers to allocate healthcare resources, construct new healthcare facilities, and develop health-related policies. For healthcare facilities, it is necessary to consider the influence of travel time across the transportation network, hospital scales, and urban population distribution to evaluate spatial accessibility and healthcare service accessibility accurately. Different methods have been proposed to evaluate hospital accessibility at city level and country level under normal conditions (Brabyn & Skelly 2002; Ngamini & Vanasse 2012; Zheng et al. 2019; De Mello-Sampayo 2018; Freiria et al. 2019; Dejen et al. 2019; Qian et al. 2020; Ghorbanzadeh et al. 2020). However, there have been relatively few studies on healthcare service accessibility after extreme events (e.g., earthquakes). Some researchers have considered the effects of post-earthquake seismic damage to buildings and transportation networks on healthcare service accessibility; for example, it was reported that 97% of earthquake-related injuries occur immediately after or within the first 30 min after the main shock (Gunn 1995). Thus, reducing the waiting time before receiving medical care is crucial for people injured in earthquakes.

The post-earthquake traffic capacity of the urban transportation network is very important in terms of a city's ability to conduct an effective emergency response and also significantly influences its healthcare service accessibility. Goretti and Sarli (2006) studied the probability of road blockage considering the effects of seismic damage to the transportation network. The post-earthquake connectivity of hospitals in the emergency response phase was analyzed. The results indicated that limited seismic damage under small earthquake intensities is negligible but the collapse of buildings along roadways under large earthquake intensities seriously influences hospital connectivity.

Ertugay et al. (2016) proposed a probabilistic estimation method for modeling post-earthquake health service accessibility. Road closure probabilities were evaluated using the average value of 10,000 Monte Carlo simulations. However, this accessibility evaluation only focused on geographic accessibility without consideration of the population of the analyzed districts, hospital size, or number of available medical staff. Umma and Luc (2017) integrated the effects of debris from damaged buildings with seismic damage to bridges and roads to evaluate post-earthquake travel ability and the systemic vulnerability of hospitals. The results indicated that post-earthquake emergency service accessibility is significantly reduced compared with normal conditions.

Previous studies on healthcare service accessibility have mostly focused on normal conditions without considering the effects of disasters such as earthquakes. Some have considered post-earthquake geographical accessibility to hospitals but failed to account for the effects of demand and hospital size. This study addresses gaps in the literature by providing a framework that accounts for the seismic damage to buildings and transportation networks, population distribution information, hospital size, and their effects on post-earthquake healthcare service accessibility. The functionality index (or functionality measure) is a parameter that can be used to describe the effectiveness of a system (e.g., hospital network) and is widely used in post-earthquake performance analysis. Functionality measures such as patient waiting time, number of available operating rooms, and economic loss caused by earthquakes are typically used for the post-earthquake performance assessment of single hospitals (Shang et al. 2020a, b).

This study focuses on an urban-level hospital network. The number of medical staff and patient beds is used to define the healthcare service in order to consider the effects of population distribution and hospital size. A healthcare service accessibility analysis framework for normal conditions and post-earthquake conditions is proposed. The effects of post-earthquake building collapse and bridge damage were integrated in evaluating the capacity of the transportation network and analyzing healthcare service accessibility. The emergency rescue demand for post-earthquake injuries and the normal demand for healthcare service for citizens in daily life were evaluated by comparison with normal conditions. Hospital site selection can be optimized according to such statistical effects of earthquakes, for example, by integrating location problems such as p-median problems (Lorena et al. 2004; Mladenović et al. 2007), maximal coverage location problems (Batta et al. 1989; Berman & Krass 2002), and allocation models (Chu & Chu 2000; Güneş & Yaman 2014) into the proposed framework.

8.2 EVALUATION FRAMEWORK AND METHOD FOR HEALTHCARE SERVICE ACCESSIBILITY

8.2.1 EVALUATION FRAMEWORK

The proposed evaluation framework for healthcare service accessibility is shown in Figure 8.1. The framework has three parts: a data acquisition module, travel cost calculation module, and accessibility measurement module. Spatial data of the transportation network, the geographical boundaries of the analyzed area, and nonspatial data of demand points (population distribution data) and service supply points (hospital distribution data) are collected first. Then, the travel cost (i.e., travel time or travel distance) between demand and supply points can be calculated using spatial analysis tools such as ArcGIS. Based on the region size of the analyzed area, the travel time threshold and decay function can be determined for further accessibility evaluation. For the healthcare service considered in this study, the selected hospital service capacity indicators include number of available patient beds and number of available medical staff. These data can be processed using methods such as 2SFCA for the subsequent analysis of spatial disparities in healthcare service accessibility.

8.2.2 TWO-STEP FLOATING CATCHMENT AREA METHOD AND ITS EXTENSION

The 2SFCA method has been widely used in measuring the spatial accessibility of healthcare service since first proposed by Radke and Mu (2000) and improved by Luo and Wang (2003). The 2SFCA method repeats the floating catchment process twice, typically once on the supply side and again on the demand side. The interaction among supply, potential demand, and travel cost is incorporated in the 2SFCA method based on the concept of the potential model.

In the first step, for each hospital location j, search all demand locations that are within a threshold travel cost from hospital j (catchment area j) and compute the supply-to-demand ratio R_j, within the catchment area using Equation (8.1). Then, for

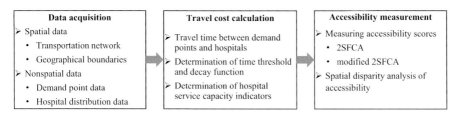

FIGURE 8.1   Evaluation framework of hospital accessibility.

each demand location i, search all hospital locations (j) within the threshold travel cost (d_0) from demand location i, and sum up the supply-to-demand ratio R_j at these locations (Equation (8.2)).

$$R_j = \frac{S_j}{\sum_{k \in \{d_{kj} \le d_0\}} D_k} \tag{8.1}$$

where D_k is the population of demand location k whose centroid falls within the catchment (that is, $d_{kj} \le d_0$), S_j is the capacity of hospital j, which is usually represented by the number of available medical staff or available patient beds, and d_{kj} is the travel cost between demand location k and hospital j.

$$A_i^F = \sum_{j \in \{d_{ij} \le d_0\}} R_j = \sum_{j \in \{d_{ij} \le d_0\}} \frac{S_j}{\sum_{k \in \{d_{kj} \le d_0\}} D_k} \tag{8.2}$$

where A_i^F represents the accessibility at demand location i based on the 2SFCA method, R_j is the supply-to-demand ratio at hospital j whose centroid falls within the catchment centered at i (that is, $d_{ij} \le d_0$), and d_{ij} is the travel cost between i and j.

The original 2SFCA method assumes that all citizens have identical accessibility within the same catchment and that the distance decay effect is roughly disposed of as 0 (within the catchment) and 1 (outside the catchment). To make this hypothesis more realistic, different methods including the enhanced 2SFCA method (E2SFCA) (Luo & Qi 2009), the Gravity 2SFCA method (G2SFCA) (Wang & Tang 2013), the Kernel Density 2SFCA method (KD2SFCA) (Dai & Wang 2011), and the Gaussian 2SFCA method (Ga2SFCA) (Dai 2010) were developed based on 2SFCA. The accessibility calculated using different modified 2SFCA methods can be summarized as Equation (8.3), where the difference lies in the distance decay function (Wang 2012). The above-mentioned methods have different assumptions for conceptualizing the distance decay, as illustrated in Equation (8.4) and Figure 8.2 (Tao & Cheng 2016).

$$A_i^F = \sum_{j \in \{d_{ij} \le d_0\}} R_j = \sum_{j \in \{d_{ij} \le d_0\}} \frac{S_j f(d_{ij})}{\sum_{k \in \{d_{kj} \le d_0\}} D_k f(d_{ij})} \tag{8.3}$$

where $f(d_{ij})$ is the generalized distance decay function based on the distance between i and j, as shown in Equation (8.3).

$$f(d_{ij}) = \begin{cases} g(d_{ij}), & d_{ij} \le d_0 \\ 0, & d_{ij} > d_0 \end{cases} \tag{3}$$

where $g(d_{ij})$ is the distance decay function within the catchment ($d_{ij} \le d_0$).

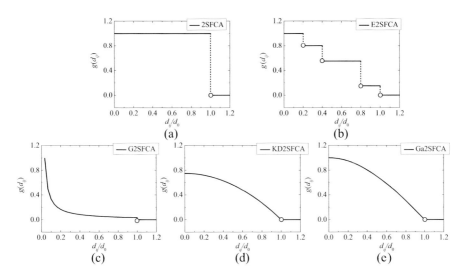

FIGURE 8.2 Major forms of distance decay function: (a) 2SFCA-binary discrete, (b) E2SFCA-multiple discrete, (c) G2SFCA-gravity function, (d) KD2SFCA-kernel density, and (e) Ga2SFCA-Gaussian function. Adapted from Tao and Cheng (2016).

For the 2SFCA method,

$$g(d_{ij}) = 1.0, \ d_{ij} \leq d_o \tag{4-1}$$

For the E2SFCA method (Luo & Qi 2009),

$$g(d_{ij}) = \begin{cases} W_1, & d_{ij} \in D_1 \\ \cdots, & \cdots \\ W_r, & d_{ij} \in D_r \end{cases} \tag{4-2}$$

where W_i ($i=1, 2, \ldots, r$) is the distance weight for the i-th travel time zone. For the G2SFCA method (Wang & Tang 2013),

$$g(d_{ij}) = d_{ij}^{-\beta}, \ d_{ij} \leq d_o \tag{4-3}$$

where β is the distance decay parameter and the value of β is 1.0 in this study. For the KD2SFCA method (Dai & Wang 2011),

$$g(d_{ij}) = \frac{3}{4}\left[1 - \left(\frac{d_{ij}}{d_0}\right)^2\right], \ d_{ij} \leq d_0 \tag{4-4}$$

and for the Ga2SFCA method (Dai 2010),

$$g(d_{ij}) = \frac{e^{-1/2 \times (d_{ij}/d_0)^2} - e^{-1/2}}{1 - e^{-1/2}}, \ d_{ij} \leq d_0 \tag{4-5}$$

Apart from the above-mentioned improvements to the 2SFCA method, the three-step floating catchment area method (3SFCA) (Wan et al. 2012), the variable 2SFCA method (V2SFCA) (Luo & Whippo 2012), and the dynamic 2SFCA method (D2SFCA) (Mcgrail & Humphreys 2014) have also been used for accessibility evaluation.

8.3 HEALTHCARE SERVICE ACCESSIBILITY UNDER NORMAL CONDITIONS BEFORE EARTHQUAKES

Recently, Shang et al. (2020c) used a Chinese medium-sized city as testbed for resilience assessment. As a primary study of hospital resilience in this testbed, healthcare service accessibility under normal conditions and post-earthquake conditions was analyzed using modified 2SFCA methods such as G2SFCA, KD2SFCA, and Ga2SFCA in this study. Detailed information on the case city can be found in Chapter 7.

8.3.1 DETERMINATION OF TRAVEL TIME THRESHOLD

Travel costs such as time and distance from demand points to hospitals are typically used in accessibility assessments (Wang 2012). To minimize the time a patient waits before receiving necessary medical care, travel time rather than travel distance is used in the calculation of spatial accessibility in this study. The minimum travel time from a demand point to a nearest hospital is used to determine the service area of the eight hospitals in the case city (Figure 8.3). Each demand point can be covered by at least one hospital in eight minutes (7.64 min).

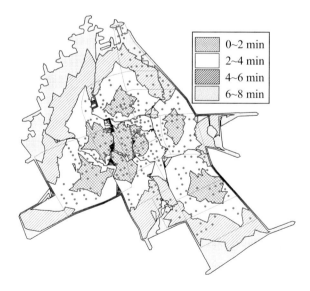

FIGURE 8.3   Service area and travel time of hospitals before earthquake.

The travel time threshold is an important determinant for accessibility assessment. The relationships between travel time from demand points to hospitals and the cumulative percentage of demand points are plotted in Figure 8.4. The minimum travel time (demand point to nearest hospital, 343 data points), maximum travel time (demand point to farthest hospital, 343 data points), and all travel time (demand points to all eight hospitals, 343×8 data points) are also shown. The travel time of 70%, 80%, and 90% of the demand points to hospitals was collected from the figure as listed in Table 8.1. Intervals of 5, 10, and 15 min were selected as travel time thresholds for sensitivity analysis. There are 39 demand points (11.37%) that cannot reach to hospitals in time if a 5-min travel time threshold is considered, which means the accessibility value is zero for these points. The population number of citizens with zero accessibility to hospitals is 107,806, accounting for 15.83% of the total population.

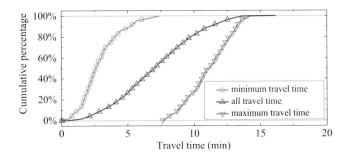

FIGURE 8.4 Cumulative percentage of citizen access to hospitals within certain travel time intervals.

TABLE 8.1
Travel Time Determined by Cumulative Percentage of Demand Points

	Cumulative Percentage	Travel Time (min)
Minimum travel time	70%	3.35
	80%	3.97
	90%	5.24
Maximum travel time	70%	12.16
	80%	12.71
	90%	13.22
All travel time	70%	8.74
	80%	9.75
	90%	11.23

8.3.2 Distance Decay Effects

The healthcare service accessibility analysis results of the 343 demand points using different distance decay functions (2SFCA, G2SFCA, KD2SFCA, and Ga2SFCA methods) and travel time thresholds (5, 10, and 15 min) are shown in Figure 8.5. Healthcare service accessibility is represented by medical staff accessibility and patient bed accessibility variables. Compared with the traditional 2SFCA, the results of the G2SFCA method tend to underestimate the healthcare service accessibility for most demand points. The accessibility for demand points with smaller travel costs is much higher than that of other demand points due to higher distance decay function $f(d_{ij})$ values as determined by the G2SFCA method (Figure 8.2c).

The underestimation and overestimation generated by the G2SFCA method are unacceptable for accessibility evaluation. If we consider the commonly used 1:3,500 physician-to-population ratio as a standard (Luo & Qi 2009), the 2SFCA and G2SFCA methods identify 107,806 persons (accounting for 15.83% of the total population and belonging to 39 demand points) without sufficient access to medical staff in the city, whereas 116,176 persons (17.06% of the total population over 42 demand points) are identified as without sufficient access to medical staff by the KD2SFCA and Ga2SFCA methods using a 5-min travel time threshold. The 1:3,500 standard is achieved per all three methods when a 10- or 15-min travel time threshold is utilized. As shown in Figure 8.5, the accessibility assessment results using KD2SFCA and Ga2SFCA methods are acceptable. Therefore, the Ga2SFCA method was used for subsequent accessibility assessments in this work.

8.3.3 Healthcare Service Accessibility and Spatial Disparity

The accessibility analysis results for medical staff and patient beds using the Ga2SFCA method are shown in Figure 8.6. The healthcare service accessibility results for demand points in central urban areas are much stronger than those in peripheral urban areas. The dense road network in the center of the city makes it easier for those citizens to travel to hospitals (or for medical staff to travel to demand points) under normal conditions. The citizens in central urban areas also have more hospitals to choose from for healthcare service under a given travel time threshold. The total population of the central urban area is also smaller than that of the peripheral urban areas and has much higher average accessibility of healthcare service.

Figure 8.6 also shows that this high accessibility in central urban areas with low accessibility in peripheral urban areas levels off with an increase in the travel time threshold. Healthcare service accessibility appears to become more equitable between peripheral and central urban areas as the travel time threshold increases. A larger travel time threshold gives citizens more hospitals to choose from over a relatively uniform distribution of healthcare resources. However, the per capita healthcare resource availability decreases as the travel time threshold increases, as shown in Figure 8.6a–c.

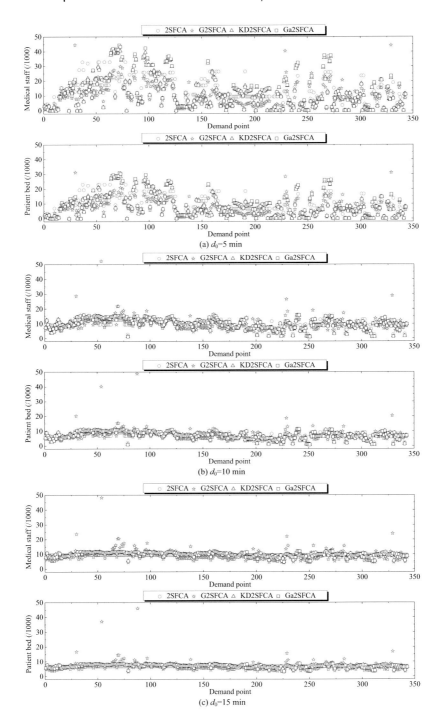

FIGURE 8.5 Comparison of accessibility of medical staff and patient beds using different decay functions and travel time thresholds.

FIGURE 8.6   Healthcare accessibility under normal conditions.

8.4 POST-EARTHQUAKE HEALTHCARE SERVICE ACCESSIBILITY

8.4.1 Post-Earthquake Healthcare Service Accessibility Analysis Framework

Seismic damage to the transportation network and debris from collapsed buildings seriously affect post-earthquake travel time across the transportation network and further affect healthcare service accessibility. Post-earthquake urban traffic capacity is affected by road damage, bridge damage, debris from collapsed buildings, and other factors. The probability of subgrade damage and road damage due to seismic activity is relatively low in a plain city (Du et al. 2018). Thus, the influence of seismic damage to the road itself was ignored in this study.

An analysis framework for post-earthquake healthcare service accessibility was developed in this study (Figure 8.7). An earthquake scenario is generated for the analyzed city and the distribution of seismic intensities in the city is calculated accordingly. The post-earthquake damage and collapse of buildings is determined based on seismic fragility analysis. The distribution of casualties as demand points for emergency rescue is calculated as well. Seismic damage to hospitals, including structures and non-structural components, is also determined. Loss of available medical staff due to injury and damage to patient beds are calculated as they significantly influence hospital functionality.

Post-earthquake damage to roads and bridges is determined under the considered earthquake scenario as indicators relevant to the transportation system. Seismic damage to the transportation network and the effects of debris from collapsed buildings are integrated to determine post-earthquake travel speed and travel time. The modified two-step floating catchment area method is used to calculate post-earthquake healthcare service accessibility. Detailed information regarding the calculation process is given below.

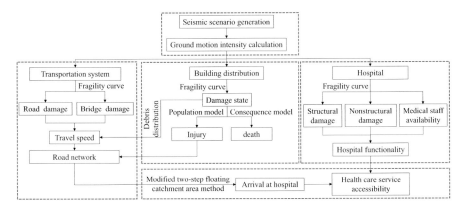

FIGURE 8.7 Post-earthquake healthcare service accessibility analysis framework.

8.4.2 Earthquake Intensity Attenuation Model

Earthquake intensity attenuation was incorporated into the seismic scenario generation in this study. The earthquake intensity is attenuated as it travels away from the epicenter. Attenuation models are often developed using data from a particular earthquake and depending on the local geology. Many models exist; selecting an appropriate model for a particular study area can be quite difficult if there is insufficient data. The attenuation model (Equation 8.5) proposed by Kawashima et al. (1984) was adopted here to simplify the seismic scenario generation process and demonstrate the feasibility of the proposed framework.

The attenuation model in Equation (8.5) is widely used in the seismic analysis of urban infrastructure systems (Yoo et al. 2016; Klise et al. 2017). Different seismic scenarios by earthquake magnitude ($M = 6$, 7, and 8) were established and compared accordingly. The epicenter is located outside the city, as shown in Figure 8.8a. The PGA distributions of earthquake magnitudes 6, 7, and 8 are shown in Figure 8.8b–d.

$$PGA = 403.8 \times 10^{0.265M} \times (R + 30)^{-1.218} \tag{8.5}$$

where M is the earthquake magnitude (unitless), R is the distance to the epicenter (km), and PGA is the peak ground acceleration (cm/s²).

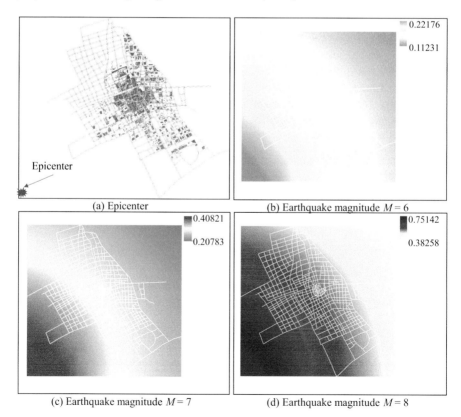

(a) Epicenter

(b) Earthquake magnitude $M = 6$

(c) Earthquake magnitude $M = 7$

(d) Earthquake magnitude $M = 8$

FIGURE 8.8 PGA distribution of earthquake magnitudes 6, 7, and 8 (unit: g).

8.4.3 Effects of Debris Induced by Building Collapse

The seismic fragility curve describes the probability that a building or bridge reaches or exceeds a damaged state given a particular engineering demand parameter (EDP) (e.g., PGA). These curves are often used to define post-earthquake damage to buildings and bridges under a given earthquake intensity. The completely damaged state (collapse) of buildings was included in this analysis. Four damage states (slight damage, moderate damage, extensive damage, and complete damage) were considered to determine post-earthquake seismic damage to bridges (Section 5.4). The seismic fragility curves shown in Figure 8.9 were used to evaluate the collapse of buildings and seismic damage to bridges in the case study city.

Based on the building distribution data and earthquake intensity data (Figure 8.8), the probability of building collapse can be calculated using seismic fragility models. There is almost no building collapse under earthquake magnitude of 6 and relatively few collapsed buildings under earthquake magnitude of 7. However, the number of collapsed buildings increases rapidly under earthquake magnitude of 8; the collapsed buildings are mainly located in central urban areas in this case, as shown in Figure 8.10.

After an earthquake, the urban transportation system suffers road congestion due to debris caused by the collapse of buildings falling along the streets. The estimation of debris extent is a complicated problem with many uncertainties as-affected by many factors. Argyroudis et al. (2015) proposed simplified geometrical models (Figure 8.11) to estimate post-earthquake debris extent. They determined the induced debris width (W_d) based on building height (Y), building width (W), the ratio (k_v) between the collapsed volume (V_T) and the original volume of the building (V_0), the inclination of the collapse (c), and the collapse mode of the building. Figure 8.11a and b show building collapse in one direction while Figure 8.11c shows building collapse in two directions. Based on the suggestions provided by Argyroudis et al. (2015) and Costa et al. (2020), the collapse mode in Figure 8.11b was adopted in this study for continuous building façades (widely used types in the case city) and a value of 45° was applied for c with a value of 50% for k_v (Argyroudis et al. 2015; Costa et al. 2020).

For each road segment, the ratio of areas not affected by debris (R) is determined by Equation (8.6-1), where W_R is the road width and W_{BR} is the distance from the

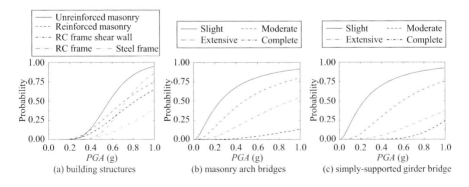

FIGURE 8.9   Fragility curves of building structures and bridges.

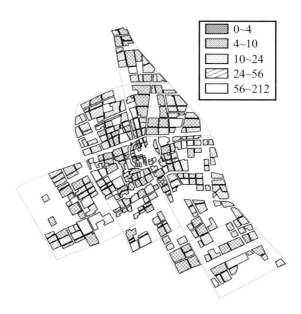

FIGURE 8.10   Average number of collapsed buildings under earthquake magnitude of 8.

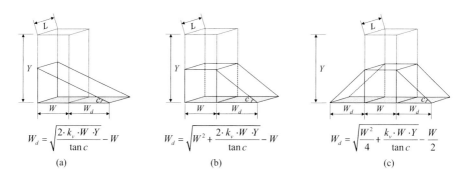

FIGURE 8.11   Estimation of debris width W_d for different collapse modes (Argyroudis et al. 2015).

edge of the building to the edge of the road segment. If R is larger than 1.0, there is no influence on travel ability across the road segment. A negative R value indicates that the road segment is impassable. When the R value ranges from 0.0 to 1.0, the post-earthquake travel speed considering the effects of collapsed buildings ($v_{\text{collapsed}}$) can be determined by Equation (8.6-2), where v_0 is the travel speed under normal conditions. If a road segment is affected by several collapsed buildings, the post-earthquake travel speed is determined by the minimum travel speed calculated considering the effects of these collapsed buildings. The average post-earthquake travel time of the considered road segment can be calculated by Equation (8.6-3), where P(collapsed) is the collapse probability of buildings, t_0 is travel time without collapsed buildings, and $t_{\text{collapsed}}$ is travel time with collapsed buildings.

$$R = \frac{W_R + W_{BR} - W_d}{W_R} \tag{8.6-1}$$

$$v_{\text{collapsed}} = R \cdot v_0 \tag{8.6-2}$$

$$t = t_{\text{collapsed}} \cdot P(\text{collapsed}) + t_0 \cdot \left(1 - P(\text{collapsed})\right) \tag{8.6-3}$$

8.4.4 EFFECTS OF BRIDGE DAMAGE

Seismic damage to bridges can be quantified by damage index DI_{bridge}, as suggested by Bocchini and Frangopol (2012). Based on seismic fragility analysis, the probability of a bridge reaching a certain damage state can be calculated by the difference between the probabilities of exceedance of damage states i and $i+1$, i.e., Equation (8.7-1). In this study, the considered damage states for a bridge were defined as "none", "slight", "moderate", "extensive", and "complete" corresponding to the seismic fragility curves shown in Figure 8.9b and c. The damage index (DI_{bridge}) of a bridge can be calculated by Equation (8.7-2). Damage to a bridge is considered minor for DI_{bridge} values less than 1.0, where there is no influence on travel ability. Bridge damage is considered moderate for DI_{bridge} values between 1.0 and 2.0, where the travel speed is half the travel speed under normal conditions. Bridge damage is considered extensive for DI_{bridge} values between 2.0 and 3.0 and complete for values greater than 3, in which cases the bridge is impassable (Umma & Luc 2017; Padgett & DesRoches 2007).

$$P(\text{no damage}) = 1 - F(\text{slight damage})$$

$$P(\text{slight damage}) = F(\text{slight damage}) - F(\text{moderate damage})$$

$$P(\text{moderate damage}) = F(\text{moderate damage}) - F(\text{extensive damage}) \tag{8.7-1}$$

$$P(\text{extensive damage}) = F(\text{extensive damage}) - F(\text{complete damage})$$

$$P(\text{complete damage}) = F(\text{complete damage})$$

$$
\begin{aligned}
DI_{bridge} = {} & 0 \cdot P(\text{no damage}) + 1 \cdot P(\text{slight damage}) + 2 \cdot P(\text{moderate damage}) \\
& + 3 \cdot P(\text{extensive damage}) + 4 \cdot P(\text{complete damage})
\end{aligned} \tag{8.7-2}
$$

where F(slight damage), F(moderate damage), F(extensive damage), and F(complete damage) are exceedance probabilities of different damage states,

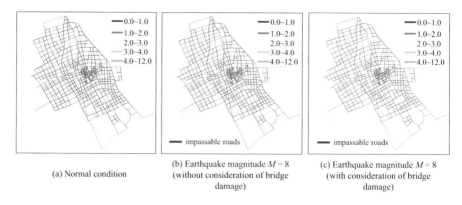

(a) Normal condition

(b) Earthquake magnitude $M = 8$ (without consideration of bridge damage)

(c) Earthquake magnitude $M = 8$ (with consideration of bridge damage)

FIGURE 8.12   Travel time before and after earthquakes.

respectively, as determined by the seismic fragility curves (Figure 8.9b and c) and PGA values of bridge locations (Figure 8.8). P(no damage), P(slight damage), P(moderate damage), P(extensive damage), and P(complete damage) are the probability of a bridge being in none, slight, moderate, extensive, or complete damage states, respectively.

8.4.5   Urban Traffic Capacity Analysis Considering Seismic Damage

The travel times of each road segment under normal conditions and under earthquake magnitude of 8 without and with consideration of bridge damage are shown in Figure 8.12a–c. The number of collapsed buildings is small under seismic scenarios with magnitudes of 6 and 7, where there is almost no influence on travel time. This is consistent with the analysis results of Goretti and Sarli (2006). Under earthquake magnitude of 8, only a few roads are impassable due to the impact of debris induced from collapsed buildings while travel time across the rest of the network is unchanged from normal conditions (Figure 8.12b). This is mainly because buildings in the case city are mainly low-rise (3 stories or fewer, 38.45%) and low multi-story buildings (4–6 stories, 36.75%). Although several buildings collapse under earthquake magnitude of 8 (Figure 8.10c), most have no influence on travel time as a result of their low height and distance from the road. The 10 masonry arch bridges and 45 simply-supported girder bridges are all in slight damage states under earthquake magnitudes of 6 and 7 and have no influence on travel time. However, for earthquake magnitude of 8, the bridges are all in moderate damage states and the travel speed falls to half of the travel speed under normal conditions (Figure 8.12c).

8.5   HEALTHCARE SERVICE ACCESSIBILITY CONSIDERING POST-EARTHQUAKE CASUALTIES

The regional post-earthquake population casualty model proposed by Yin (1991) was adopted here to determine the number of injuries and deaths. The proportion of deaths to the total population of the region (D) was determined by the proportion of collapsed buildings in the studied area (*A*) by Equation (8.8). Yin (1991) suggested that the number of injured people is generally three to five times the number of deaths. The proportion of injuries to the total population of the region (*I*) was determined as three times of the death proportion (*D*) (*I* = 3*D*) (Xiao 1987; Zhang & Jiang 2016). Each demand point was selected as the studied area when calculating injuries and deaths as shown in Figure 8.13 and Table 8.2.

The number of collapsed buildings under earthquake magnitude of 6 is limited and there are no casualties. The number of collapsed buildings under earthquake magnitude of 7 is 145 and there are 32 people injured. This number increases rapidly under earthquake magnitude of 8, where the number of collapsed buildings is almost 1,711 and there are 4,110 people injured accounting for 6.036‰ of the total population.

$$D = 10^{12.479 \cdot A^{0.1} - 13.3} \tag{8.8}$$

The demand for medical staff and patient beds in these scenarios were calculated as follows (Nie et al. 2001):

$$D_{MS} = 0.1039\omega \cdot N \tag{9-1}$$

$$D_{PB} = 0.2163\omega \cdot N \tag{9-2}$$

where *N* is the number of injured citizens, *ω* is the zone coefficient of the city (*ω* = 1.51 in this case), D_{MS} is the demand for medical staff, and D_{PB} is the demand for patient beds.

▨	0~8
▥	8~19
▢	19~36
▨	36~73
▢	73~144

FIGURE 8.13 Distribution of injured population in each demand point under earthquake magnitude of 8.

TABLE 8.2
Number of Post-Earthquake Casualties

Earthquake magnitude	6	7	8
Number of injuries	0	32	4,110
Proportion of injuries	0	4.76333E−05	0.00603588
Number of deaths	0	11	1,370
Proportion of deaths	0	1.58778E−05	0.00201196

The available number of patient beds and medical staff available for post-earthquake emergency rescue can be calculated as follows:

$$S_{PB} = a \cdot N_{PB} \cdot (1-b) \tag{10-1}$$

$$S_{MS} = N_{MS} \cdot (1-c) \cdot (1-d) \tag{10-2}$$

where N_{PB} is the total number of patient beds in a hospital, N_{MS} is the total number of medical staff in a hospital, S_{PB} and S_{MS} are the numbers of available patient beds and medical staff (supply of a hospital), a is the functionality preservation rate of urban hospitals after earthquakes, which represents the proportion of undamaged patient beds, b is the occupancy rate of hospital beds before earthquakes (selected as 76%), c is the proportion of injured and dead medical staff induced by hospital collapse, and d is the occupancy rate of medical staff under normal conditions (selected as 50%). The values of a and c can be determined by seismic fragility analysis of the hospital buildings.

The total number of injuries is 4,110 under earthquake magnitude of 8. The number of post-earthquake available medical staff and patient beds in each hospital was determined here by Equation (8.10). A total of 709 injured people (17.3% of total injuries) distributed in 48 demand points were identified as without access to healthcare service using a 5-min travel time threshold. Remaining injured people all have accessibility to healthcare service and can satisfy the demand for medical staff determined by Equation (8.10–2). However, there are 853 injured people (20.8% of total injuries) without sufficient access to healthcare if patient beds is selected as the functionality index of healthcare service. The demand points that cannot meet the requirement for healthcare service accessibility in an emergency rescue scenario were mainly distributed in the central urban area, where some roads are impassable, as well as areas on the city edge that are far from hospitals (Figure 8.14a).

When the effects of seismic damage to bridges were included in the accessibility evaluation, a total of 736 injured people (17.9% of total injuries) distributed in 64 demand points were identified as without access to healthcare service using a 5-min travel time threshold (Figure 8.14b). Remaining injured people in this case have access to healthcare service and can satisfy the demand for medical staff. There are 906 injured people (22.8% of total injuries) without sufficient access to patient beds.

The post-earthquake accessibility to healthcare service for citizens in normal daily life is described in Table 8.3, where the number of medical staff is the functionality index for healthcare service and the commonly used 1:3,500 physician-to-population

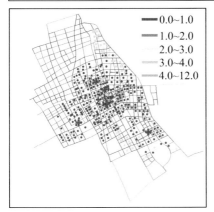
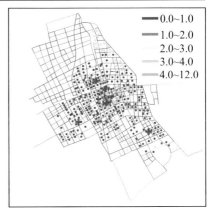

(a) without consideration of bridge damage   (b) with consideration of bridge damage

FIGURE 8.14   Demand points where accessibility does not satisfy emergency rescue requirements.

ratio is the standard (Luo & Qi 2009) Under earthquake magnitudes of 6 and 7, seismic damage does not significantly influence travel time on each road segment and the corresponding healthcare service accessibility is similar to normal conditions (Sections 4.3 and 4.4). However, the portion of people without healthcare service accessibility increases from 17.06% under normal conditions to 19.9% and 25.0% under earthquake magnitude of 8 without or with consideration of bridge damage, respectively. Large earthquakes severely influence citizens' healthcare service accessibility in the case study city, particularly when the seismic damage to bridges is taken under consideration.

8.6   SUMMARY AND CONCLUSIONS

City-level post-earthquake accessibility to healthcare service was analyzed in this study. A quantitative framework that integrates seismic fragility analysis and accessibility assessment was established and validated in a case study on a medium-sized city. The integrated framework fills current gaps in the literature by incorporating the effects of seismic damage to buildings and transportation networks on post-earthquake travel time to hospitals. Population distribution information and hospital size are also included in the evaluation of earthquake-induced casualties and hospital service capability. The number of medical staff and patient beds supplied by hospitals in the analyzed city was selected as the healthcare service functionality index.

The proposed framework was adopted to evaluate the healthcare service accessibility of a medium-sized city in China. The results under normal conditions indicated that travel time thresholds significantly influence healthcare service accessibility. The effects of different distance decay functions were also analyzed. The Ga2SFCA

TABLE 8.3

Post-Earthquake Healthcare Accessibility for Citizens in Normal Daily Life Under 5-Min Travel Time Threshold

Earthquake Magnitude	Indicators	Number	Portion
6	Demand points without access to healthcare service	44	12.8%
	Population without access to healthcare service	121,078	17.8%
	Demand points not satisfying demand for healthcare service	47	13.7%
	Population not satisfying the demand for healthcare service	124,946	18.3%
7	Demand points without access to healthcare service	46	13.4%
	Population without access to healthcare service	129,005	18.9%
	Demand points not satisfying the demand for healthcare service	49	14.3%
	Population not satisfying the demand for healthcare service	132,873	19.5%
8 (Without consideration of bridge damage)	Demand points without access to healthcare service	48	14.0%
	Population without access to healthcare service	133,968	19.7%
	Demand points not satisfying demand for healthcare service	50	14.6%
	Population not satisfying the demand for healthcare service	135,638	19.9%
8 (With consideration of bridge damage)	Demand points without access to healthcare service	64	18.7%
	Population without access to healthcare service	162,681	23.9%
	Demand points not satisfying the demand for healthcare service	69	20.1%
	Population not satisfying the demand for healthcare service	170,521	25.0%

method and a 5-min travel time threshold were selected for accessibility evaluation. The results indicated that most of the road segments are not influenced by collapsed buildings even under earthquake magnitudes as high as 8. This is because buildings in the city are mainly low-rise and are located far away from roads. However, the overall travel time does increase if seismic damage to bridges is taken into account.

Based on the travel time analysis results, post-earthquake healthcare service accessibility evaluation was conducted considering the number of injuries and deaths. The emergency rescue demand relevant to post-earthquake injuries and the normal demand for quotidian healthcare service were considered and compared between post-earthquake and normal conditions. The results indicate that the demand points that cannot satisfy the requirement of healthcare service accessibility for emergency rescue are mainly distributed in the central urban area (with impassable roads) and areas in the city edge (far from the hospitals).

The proposed framework and analysis results can provide a workable reference for the site selection of medical facilities in urban planning processes, the optimization of square cabin hospital site selections after earthquakes, and the reconstruction of urban transportation infrastructure systems. However, this work was not without limitations. In the future, the interactions between patients, ambulances, medical staff, and transportation networks can be modeled by system simulation and included in the proposed framework. The wait time from arrival at hospitals to the provision of care to patients can also be considered. Optimization theories also may be integrated into the proposed framework to further rationalize hospital site selection considering the effects of earthquakes.

REFERENCES

Argyroudis S, Selva J, Gehl P, Pitilakis K. 2015. Systemic seismic risk assessment of road networks considering interactions with the built environment, *Computer-Aided Civil and Infrastructure Engineering*, **30**(7), 524–540. doi:10.1111/mice.12136.

Batta R, Dolan J M, Krishnamurthy N N. 1989. The maximal expected covering location problem: Revisited, *Transportation Science*, **23**(4), 277–287. doi:10.1287/trsc.23.4.277.

Berman O, Krass D. 2002. The generalized maximal covering location problem, *Computers Operations Research*, **29**(6), 563–581. doi:10.1016/s0305-0548(01)00079-x.

Bocchini P, Frangopol D M. 2012. Restoration of bridge networks after an earthquake: Multicriteria intervention optimization, *Earthquake Spectra*, **28**(2), 426–455. doi:10.1193/1.4000019.

Brabyn L, Skelly C. 2002. Modeling population access to New Zealand public hospitals, *International Journal of Health Geographics,* **1**(1), 1–9.

Chu S C K, Chu L. 2000. A modeling framework for hospital location and service allocation, *International Transactions in Operational Research,* **7**(6), 539–568. doi:10.1111/j.1475-3995.2000.tb00216.x.

Costa C, Figueiredo R, Silva V, Bazzurro P et al. 2020. Application of open tools and datasets for probabilistic modeling of road traffic disruptions due to earthquake damage, *Earthquake Engineering Structural Dynamics*, **49**(12), 1236–1255. doi:10.1002/eqe.3288.

Dai D. 2010. Black residential segregation disparities in spatial access to health care facilities and late-stage breast cancer diagnosis in metropolitan Detroit, *Health Place*, **16**(5), 1038–1052. doi:10.1016/j.healthplace.2010.06.012.

Dai D, Wang F. 2011. Geographic disparities in accessibility to food stores in southwest Mississippi, *Environment and Planning B: Planning and Design*, **38**(4), 659–677. doi:10.1068/b36149.

De Mello-Sampayo F. 2018. Spatial interaction healthcare accessibility model: An application to Texas, *Applied Spatial Analysis and Policy*. doi:10.1007/s12061-018-9284-4.

Dejen A, Soni S, Semaw F. 2019. Spatial accessibility analysis of healthcare service centers in Gamo Gofa Zone Ethiopia through Geospatial technique, *Remote Sensing Applications: Society and Environment*, **13**, 466–473. doi:10.1016/j.rsase.2019.01.004.

Du J, Lin J Q, Liu J L. 2018. Research and application of recoverability evaluation method for urban road network after earthquake, *Science Discovery,* **6**(5), 327–331. doi:10.11648/j.sd.20180605.13 (in Chinese).

Dutt A, Dutta H, Jaiswal J, Monroe C. 1986. Assessment of service adequacy of primary health care physicians in a two county region of Ohio USA, *GeoJournal*, **12**(4). doi:10.1007/bf00262368.

Ertugay K, Argyroudis S, Düzgün H Ş. 2016. Accessibility modeling in earthquake case considering road closure probabilities: A case study of health and shelter service accessibility in Thessaloniki Greece, *International Journal of Disaster Risk Reduction*, **17**, 49–66. doi:10.1016/j.ijdrr.2016.03.005.

Freiria S, Tavares A O, Julio R P. 2019. The benefits of a link-based assessment of health services accessibility: Unveiling gaps in central region of Portugal, *Land Use Policy*, **87**, 104034. doi:10.1016/j.landusepol.2019.104034.

Ghorbanzadeh M, Kim K, Ozguven E E, Horner M W. 2020. A comparative analysis of transportation-based accessibility to mental health services, *Transportation Research Part D: Transport and Environment*, **81**, 102278. doi:10.1016/j.trd.2020.102278.

Goretti A, Sarli V. 2006. Road network and damaged buildings in urban areas: Short and long-term interaction, *Bulletin of Earthquake Engineering*, **4**(2), 159–175. doi:10.1007/s10518-006-9004-3.

Guagliardo M F. 2004. Spatial accessibility of primary care: Concepts methods and challenges. *International Journal of Health Geographics* **3**(1), 1–13. doi:10.1186/1476-072x-3-3.

Gunn S W A. 1995. Health effects of earthquakes, *Disaster Prevention Management,* **4**(5), 6–10. doi:10.1108/09653569510100956.

Güneş E D, Yaman H. 2014. Health network mergers and hospital re-planning, *Journal of the Operational Research Society*, 275–283. doi:10.1057/jors.2008.165.

Hansen W G. 1959. How accessibility shapes land use, *Journal of the American Institute of Planners*, **25**(2), 73–76. doi:10.1080/01944365908978307.

Kawashima K, Aizawa K, Takahashi K. 1984. Attenuation of peak ground motion and absolute acceleration response spectra. In: *Proceedings of the 8th World Conference on Earthquake Engineering* (*WCEE*), International Association for Earthquake Engineering (IAEE), Tokyo, pp. 257–264.

Klise K A, Bynum M, Moriarty D, Murray R. 2017. A software framework for assessing the resilience of drinking water systems to disasters with an example earthquake case study, *Environmental Modelling Software*, **95**, 420–431. doi:10.1016/j.envsoft.2017.06.022.

Lorena L A N, Senne E L F. 2004. A column generation approach to capacitated p-median problems, *Computers Operations Research*, **31**(6), 863–876. doi:10.1016/s0305-0548(03)00039-x.

Luo W, Wang F. 2003. Measures of spatial accessibility to health care in a gis environment: Synthesis and a case study in the Chicago region, *Environment and Planning B: Planning and Design*, **30**(6), 865–884. doi:10.1068/b29120.

Luo W, Qi Y. 2009. An enhanced two-step floating catchment area (E2SFCA) method for measuring spatial accessibility to primary care physicians, *Health Place*, **15**(4), 1100–1107. doi:10.1016/j.healthplace.2009.06.002.

Luo W, Whippo T. 2012. Variable catchment sizes for the two-step floating catchment area (2SFCA) method, *Health Place*, **18**(4), 789–795. doi:10.1016/j.healthplace.2012.04.002.

Mcgrail M R, Humphreys J S. 2014. Measuring spatial accessibility to primary health care services: Utilising dynamic catchment sizes, *Applied Geography*, **54**, 182–188. doi:10.1016/j.apgeog.2014.08.005.

Mladenović N, Brimberg J, Hansen P, Moreno-Pérez J A. 2007. The p-median problem: A survey of metaheuristic approaches, *European Journal of Operational Research*, **179**(3), 927–939. doi:10.1016/j.ejor.2005.05.034.

Ngamini Ngui A, Vanasse A. 2012. Assessing spatial accessibility to mental health facilities in an urban environment, *Spatial and Spatio-Temporal Epidemiology,* **3**(3), 195–203. doi:10.1016/j.sste.2011.11.001.

Nie G Z, Gao J G, Su G W, Wang J M. 2001. Models on rapid judgment for the emergent rescue needs during earthquake: By analysis on post-earthquake events, *Resources Science*, **23**(1), 69–76 (in Chinese).

Padgett J E, DesRoches R. 2007. Retrofitted bridge fragility analysis for typical classes of multi-span bridges, *Earthquake Spectra*, **23**(1), 115–130. doi:10.1193/1.3049405.

Qian T, Chen J, Li A, Wang J, Shen D. 2020. Evaluating spatial accessibility to general hospitals with navigation and social media location data: A case study in Nanjing, *International Journal of Environmental Research and Public Health*, **17**(8), 2752. doi:10.3390/ijerph17082752.

Radke J, Mu L. 2000. Spatial decompositions modeling and mapping service regions to predict access to social programs, *Annals of GIS*, **6**(2), 105–112. doi:10.1080/10824000009480538.

Shang Q X, Wang T, Li J C. **2020a.** Seismic resilience assessment of emergency departments based on the state tree method, *Structural Safety*, **85**, 101944. doi:10.1016/j.strusafe.2020.101944.

Shang Q. X, Wang T, Li J. C. 2020b. A quantitative framework to evaluate the seismic resilience of hospital systems, *Journal of Earthquake Engineering.* doi:10.1080/13632469.2020.1802371.

Shang Q, Guo X, Li Q, Xu Z, Xie L, Liu C, Li J, Wang T. 2020c. A benchmark city for seismic resilience assessment, *Earthquake Engineering and Engineering Vibration*, **19**(4), 811–826. doi:10.1007/s11803-020-0597-3.

Spencer J, Angeles G. 2007. Kernel density estimation as a technique for assessing availability of health services in Nicaragua, *Health Services and Outcomes Research Methodology*, **7(3–4)**, 145–157. doi:10.1007/s10742-007-0022-7.

Tao Z L, Cheng Y. 2016. Research progress of the two-step floating catchment area method and extensions, *Progress in Geography*, **35**(5), 589–599. doi:10.18306/dlkxjz.2016.05.006.

Umma T, Luc C. 2017. Systemic seismic vulnerability of transportation networks and emergency facilities, *Journal of Infrastructure Systems*, **23**(4), 04017032. doi:10.1061/(ASCE)IS.1943-555X.0000392.

Wan N, Zou B, Sternberg T. 2012. A three-step floating catchment area method for analyzing spatial access to health services, *International Journal of Geographical Information Science*, **26**(6), 1073–1089. doi:10.1080/13658816.2011.624987.

Wang F. 2012. Measurement optimization and impact of health care accessibility: A methodological review, *Annals of the Association of American Geographers*, **102**(5), 1104–1112. doi:10.1080/00045608.2012.657146.

Wang F, Tang Q. 2013. Planning toward equal accessibility to services: A quadratic programming approach, *Environment and Planning B: Planning and Design*, **40**(2), 195–212. doi:10.1068/b37096.

WHO. 2005. *WHO Sustainable Health Financing Universal Coverage and Social Health Insurance.* https://www.who.int/health_financing/documents/cov-wharesolution5833/en/ (accessed on 1st August, 2020).

Yin Z Q. 1991. A study for predicting earthquake disaster loss, *Earthquake Engineering and Engineering Vibration*, **11**(4), 87–96 (in Chinese).

Xiao G X. 1987. A prediction method for earthquake loss, *Journal of Seismology*, 1, 1–8 (in Chinese.

Yoo D G, Jung D, Kang D, Kim J H, Lansey K. 2016. Seismic hazard assessment model for urban water supply networks, *Journal of Water Resources Planning and Management*, **142**(2), 04015055. doi:10.1061/(asce)wr.1943-5452.0000584.

Zhang W L, Jiang H J. 2016. A review of methods and models on seismic casualty estimation, *Structural Engineers*, **32**(3), 181–191. doi:10.15935/j.cnki.jggcs.2016.03.027 (in Chinese).

Zheng Z, Xia H, Ambinakudige S, Qin Y, Li Y, Xie Z, Zhang L, Gu H. 2019. Spatial accessibility to hospitals based on web mapping API: An empirical study in Kaifeng China, *Sustainability*, **11**(4), 1160. doi:10.3390/su11041160.

9 Summary and Remarks

9.1 MAJOR ACHIEVEMENTS AND CONTRIBUTIONS

This monograph systematically presents a suite of novel techniques developed by the authors and their team for seismic resilience assessment of hospital infrastructure, with particular emphasis on seismic tests and fragility models of hospital equipment, resilience assessment of single hospital buildings and emergency departments, and post-earthquake functionality of urban hospital infrastructures.

Chapters 2–4 focus on seismic performance and fragility of hospital nonstructural components including medical equipment. Chapter 2 describes seismic force demands on acceleration-sensitive nonstructural components. Different methods for generating floor response spectra (FRS) are presented first, followed by a review of the amplification factor methods. Detailed investigations of research on the critical factors affecting FRS are outlined. The floor acceleration response and FRS obtained from experimental studies and field observations during earthquakes are discussed. The major knowledge gaps to be filled are identified, and possible future research challenges are discussed. Chapter 3 proposed a simplified method based on the modification of an existing methodology for the generation of FRS. The results calculated by time history analyses are used to validate the proposed amplification factors and demonstrate the feasibility and effectiveness of the method. FRS for seismic performance tests of NSCs were established based on the analysis results. The proposed response spectrum is composed of short period section, linear increase section, platform section, and decrease section. Chapter 4 conducted shaking table tests of different types of medical equipment including infusion supports, medical cabinets, medical resuscitation carts, patient beds, and shadowless lamps using the proposed response spectrum. Seismic response and damage states of these medical equipment were analyzed. The results indicated that seismic damages of infusion supports and medical cabinets are mainly rocking and overturning, while the response of medical resuscitation carts, patient beds, and shadowless lamps is mainly controlled by sliding and rolling. Then, seismic fragility curves of the tested medical equipment were generated based on the test results.

Chapters 5 and 6 focus on single hospital building level resilience assessment. Chapter 5 proposed a quantitative framework to evaluate the seismic resilience of single hospital buildings. A typical hospital system is categorized, and importance factors for different functional units, subsystems, and components are quantified. Resilience demand is expressed as the desirable recovery time of the hospital system after earthquakes. Seismic resilience is quantified based on probabilistic seismic fragility analysis. Recovery time is calculated considering an idealized repair path. Loss of functionality of the hospital is evaluated as the sum of weighted economic losses of all components. Chapter 6 developed a new method called the state tree method that explicitly considers the component contribution to the functionality of

DOI: 10.1201/9781003457459-9

the emergency department. A typical emergency department is analyzed, and the state tree model is developed. The component contribution to the functionality is explicitly expressed using the fault tree method, and the failure transfer mechanism is determined by success paths. The functionality is thus defined as the ratio of successful paths to total possible paths. The system fragility is then calculated based on the full probability theory, and the results are used to validate the simulation method. The critical components that significantly affect the functionality and the recovery can be identified while considering a practical repair process. Finally, the resilience curve of the emergency department is developed based on the fragility of the components.

Chapters 7 and 8 focus on network level seismic resilience assessment of hospital infrastructure systems. Chapter 7 developed a Geographic Information System (GIS)–based benchmark model of a medium-sized city located in the southeastern coastal region of China. The benchmark city can be used to compare existing assessment frameworks and calibrate the assessment results. The demographics, site conditions, and potential hazard exposure of the benchmark city, as well as land use and building inventory, are described in this chapter. Data on lifeline systems are provided, including power, transportation, water, drainage, and natural gas distribution networks, as well as the locations of hospitals, emergency shelters, and schools. Data from past earthquakes and the literature are obtained to develop seismic fragility models, consequence models, and recovery models, which can be used as basic data or calibration data in the resilience assessment process. Chapter 8 presents a quantitative framework that considers seismic damage to buildings and transportation networks, and post-earthquake available numbers of patient beds and medical staff to evaluate post-earthquake healthcare service accessibility. The modified two-step floating catchment area method and seismic fragility analysis are integrated into the proposed framework. Spatial distributions of demand point, hospital, and transportation network data are used to evaluate healthcare service accessibility.

9.2   A FUTURE PERSPECTIVE

The seismic resilience assessment of the hospital infrastructure system involves the cross-integration of multiple disciplines. This work focuses on seismic fragility analysis at component level, seismic resilience evaluation of single medical building, and post-earthquake accessibility calculation of urban medical services. The developed models and frameworks proposed in this work provide useful information to guide seismic performance analysis of hospital infrastructure. In addition, it should be noted that the post-earthquake emergency response and recovery of the hospital infrastructure system are closely related to the seismic damage and recovery process of various urban infrastructure systems, including the power, transportation, water, and natural gas distribution networks. Only the interactions between the hospitals, urban buildings and transportation networks are integrated in the analysis of this work. The influence of the interdependence between these infrastructure systems should be carefully considered in further research.

Index

Note: **Bold** page numbers refer to tables; *italic* page numbers refer to figures.